Robert Palmer

Silicon Photonic Modulators for Low-power Applications

Karlsruhe Series in Photonics & Communications, Vol. 18
Edited by Profs. J. Leuthold, W. Freude and C. Koos

Karlsruhe Institute of Technology (KIT)
Institute of Photonics and Quantum Electronics (IPQ)
Germany

Silicon Photonic Modulators for Low-power Applications

by

Robert Palmer

KIT Scientific Publishing

Dissertation, Karlsruher Institut für Technologie (KIT)
Fakultät für Elektrotechnik und Informationstechnik, 2014

Impressum

KIT Scientific Publishing

Karlsruher Institut für Technologie (KIT)
KIT Scientific Publishing
Straße am Forum 2
D-76131 Karlsruhe

KIT Scientific Publishing is a registered trademark of Karlsruhe Institute of Technology. Reprint using the book cover is not allowed.

www.ksp.kit.edu

Print on Demand 2015

ISSN 1865-1100
ISBN 978-3-7315-0386-6
DOI 10.5445/KSP/1000047154

Silicon Photonic Modulators for Low-power Applications

Zur Erlangung des akademischen Grades eines

DOKTOR-INGENIEURS

von der Fakultät für Elektrotechnik und Informationstechnik
des Karlsruher Instituts für Technologie (KIT)

genehmigte

DISSERTATION

von

Dipl.-Phys. Robert Palmer

geboren in

Pawlodar, Kasachstan

Tag der mündlichen Prüfung: 22.05.2014
Hauptreferent: Prof. Dr. sc. nat. Juerg Leuthold
Korreferent: Prof. Dr.-Ing. Dr. h. c. Wolfgang Freude

Table of Contents

Abstract (German)

Die vorliegende Arbeit behandelt neue Konzepte zur Herstellung und Funktionalisierung integrierter optischer Schaltkreise aus Silizium. Ziel dieser Untersuchung ist die Prüfung der Konzepte auf Eignung zur Reduktion des Energieverbrauchs optischer Schaltkreise im Vergleich zu existierenden Ansätzen der Siliziumtechnologie. Konkret wird ein nass-chemisches Siliziumätzverfahren entwickelt, welches das Potential hat, sowohl niedrige Wellenausbreitungsverluste auf dem Chip als auch niedrige Einkoppelverluste in den Chip zu ermöglichen. Weiterhin wird ein Silizium-Organik Hybrid(SOH)-Ansatz untersucht, welcher die optischen Eigenschaften von Silizium mit denen elektrooptisch aktiver organischer Materialien kombiniert und damit die Herstellung von integrierten elektrooptischen Modulatoren mit nie dagewesener Modulationseffizienz ermöglicht. Auf die theoretische Erarbeitung der Konzepte und auf die Untersuchung ihres Potentials folgt die Charakterisierung hergestellter Prototypen sowie deren Einsatz in Datenübertragungsexperimenten.

Der exponentielle Anstieg des globalen Internetdatenverkehrs bringt steigende Anforderungen mit sich sowohl im Hinblick auf die benötigte Übertragungskapazität als auch bezogen auf die Reduktion des zugehörigen Energieverbrauchs. Ein Kernstück der Kommunikationsinfrastruktur sind gewaltige Datenzentren von global agierenden Konzernen wie Google, Microsoft oder Facebook. So hat beispielsweise das jüngst in Luleå, Schweden, gebaute Datenzentrum von Facebook eine Fläche von 27.000 m^2 und einen Energieverbrauch von 120 MW [1]. Dabei wurde das Datenzentrum bereits strategisch günstig nahe am Polarkreis platziert, um den benötigten Energiebedarf für Kühlsysteme auf ein Minimum zu reduzieren.

Analysen haben ergeben, dass drei Viertel des Datenverkehrs eines Datenzentrums auf die Kommunikation zwischen einzelnen Servern im Datenzentrum selbst entfallen [2]. Diese Intra-Kommunikationsnetzwerke werden schon heute durch optische Datenübertragungsstrecken realisiert. Stand der Technik stellen dabei so genannte „optische Kabel", basierend auf direktmodulierten, vertikal emittierenden Lasern dar (*engl.* vertical cavity surface emitting laser, VCSEL). Im Hinblick auf steigende Datenraten und die erforderliche Reduktion des Energieverbrauchs versucht man, diese direkt modulierten Laser durch Dauerstrichlaser mit integrierten elektro-optischen Modulatoren abzulösen. Ziel ist es hierbei, den Energieverbrauch zu reduzieren, große Datenraten bei hoher Signalqualität zu übertragen und die Bauteile in optisch-integrierten Schaltkreisen zu realisieren. Dabei muss der Stückpreis niedrig gehalten werden, wobei die Signalqualität mit der von diskreten Realisierungen vergleichbar bleiben muss.

Silizium-photonische Schaltungen auf Silicon-on-Insulator-Substraten (SOI) versprechen kostengünstige Transceiver-Module, welche den oben genannten Ansprüchen genügen. Silizium zeichnet sich hier durch viele positive Eigenschaften aus, wie z. B. Transparenz im nahen Infraroten, was für den Telekommunikationssektor von großer Bedeutung ist. Die hohe Brechzahl von Silizium ($n = 3.5$) gestattet kompakte Wellenleitergeometrien und damit eine hohe Integrationsdichte. Ein typischer Wellenleiter ist lediglich 220 nm hoch und 450 nm breit. Die für die Fabrikation nötige Technologie ist größtenteils identisch mit der aus der integrierten Elektronik stammenden CMOS-Technologie, so dass bereits eine massenfabrikationstaugliche Fertigungsinfrastruktur für die Silizium-Photonik verfügbar ist.

In dieser Arbeit werden neue Fabrikationskonzepte und Technologien erarbeitet und untersucht, welche teilweise Prozesse der CMOS-Technologie ersetzen und teilweise, auf CMOS-Fabrikation aufbauend, deren Prozesskette erweitern.

Kapitel 1 beschreibt den Stand der Technik der Silizium-Photonik in den Bereichen Fabrikation, Anbindung von Glasfasern an Silizium-Chips und elektrooptische Modulatoren.

Kapitel 2 vermittelt das theoretische und technologische Hintergrundwissen. Das Kapitel umfasst eine Beschreibung nichtlinearer optischer Effekte zweiter Ordnung, insbesondere in organischen Materialien, einen Überblick über gebräuchliche Wellenleitertypen der Silizium-Photonik, die Funktionsweise zweier gebräuchlicher Strukturen zur effizienten Ankopplung von Glasfasern an Siliziumchips sowie die Grundlagen und das Konzept eines silizium-organischen Hybrid-Modulators.

Kapitel 3 befasst sich mit einem neuartigen Fertigungskonzept für integrierte Siliziumwellenleiterstrukturen, basierend auf einem anisotropen nass-chemischen Ätzprozess. Wir demonstrieren erste hergestellte Strukturen in Form von Wellenleitern und Gitterkopplern mit trapezoidaler Querschnittsfläche. Die Strukturen weisen eine effektive Seitenwandrauigkeit von 2 nm auf [C5]. Des Weiteren zeigen wir, dass es bei Wahl eines geeigneten Substrates möglich ist, Gitterkoppler herzustellen mit Einfügeverlusten von 0.3 dB zwischen Faser und Chip. Der nass-chemische Prozess erfordert die präzise Ausrichtung aller Wellenleitergeometrien nach bestimmten Kristallebenen von

Silizium. Diese Bedingung schränkt das Anwendungsgebiet insoweit ein, als dass nur gerade Wellenleiter gefertigt werden können. Für gekrümmte Strukturen ist eine Kombination von Trockenätz- und Nass-Ätzprozessen unausweichlich. Hierzu untersuchen wir die Einfügeverluste zwischen verschiedenen Wellenleitergeometrien, insbesondere hinsichtlich möglicher Ausrichtungsungenauigkeiten zwischen den zugehörigen lithographischen Ebenen. Zuletzt demonstrieren wir die Möglichkeit, durch ein spezielles Maskenlayout Strukturen herzustellen, deren Dimensionen die Auflösung des verwendeten lithographischen Verfahrens deutlich unterschreiten [C5].

In Kapitel 4 wird ein Konzept zur Faser-Chip-Kopplung behandelt, bei dem die Faserachse parallel zur Chip-Fläche verläuft. Ein Schrägschliff der Faserendfläche unter 40° lenkt das in der Faser geführte Licht auf einen Standard-Gitterkoppler auf dem Siliziumchip. Wir untersuchen dieses Konzept mit Simulationen und Experimenten auf charakteristische Kenngrößen, wie Kopplungsgrad und Ausrichtungstoleranz. Zuletzt untersuchen wir, wie weit sich der Kopplungsgrad verbessern lässt, wenn man das Gitter an das Profil des Strahls anpasst, der nach Totalreflexion an der schrägen Faserendfläche aus dem Fasermantel austritt.

In Kapitel 5 behandeln wir energieeffiziente elektrooptische Modulatoren, gefertigt in silizium-organischer Hybrid-Technologie. Wir beschreiben das Konzept eines Gegentakt-Mach-Zehnder-Modulators, den Entwurf und die Charakterisierung der verwendeten integrierten passiv-optischen Bauelemente und den Einsatz verschiedener elektrooptisch aktiver Materialien sowie deren Polung [C30]. Wir entwerfen und charakterisieren Koplanarelektroden für Frequenzen bis 50 GHz, bestimmen den Frequenzgang der Modulatoren und vergleichen die Messungen mit einem analytischen Modell. Wir überprüfen die Bauteile in Systemexperimenten unter Verwendung unterschiedlicher Modulationsformate bei Datenraten bis zu 84 Gbit/s [C21, J5]. Ferner demonstrieren wir On-off-keying bei einer Datenrate von 12.5 Gbit/s mit einem Energieverbrauch unter 1 fJ/bit [C28, J15].

Kapitel 6 fasst die wesentlichen Ergebnisse dieser Arbeit zusammen und gibt einen Ausblick über die künftige Entwicklung bei der Fabrikation von Silizium-Chips, der Faser-Chip-Kopplung und der Energieeffizienz von Siliziummodulatoren.

Preface

Global internet traffic appears to continue its exponential growth, calling for significant advancements of communication infrastructure and technology to increase transmission capacity, while simultaneously reducing the associated energy consumption. In information and communication technology the biggest source of energy consumption are large data centers. The power consumption of the data centers of global operating organizations like Google, Microsoft or Facebook is rapidly approaching the value of 1 GW per data center [1].

A traffic analysis of data centers [2] shows that almost 75 % of the traffic is due to communication between servers within the data center. Therefore the advancement of communication links between the servers offers a large potential to reduce the energy consumption of the data center. Already today these intra-communication networks rely on optical links. State-of-the-art technology are so-called optical cables with directly modulated vertical cavity surface emitting lasers (VCSEL). In the near future, these directly modulated lasers will be replaced by continuous wave (CW) lasers with high-speed integrated electro-optic (EO) modulators for a larger link capacity. In addition, these EO modulators need to feature low energy consumption, small footprint, and low cost, while the resulting signal quality has to be comparable to those of transmission links that are composed of discrete components.

Photonic integrated circuits (PIC) in silicon are a promising candidate to fulfil these demands. Silicon photonics has a high integration density resulting in a small footprint of the PICs. Furthermore, fabrication of silicon PICs mainly relies on mature CMOS processes, offering the possibility of large-scale mass production at reduced unit cost. However, prior to using silicon PICs in industry, a number of challenges need to be addressed, such as realizing a light source on-chip, a low optical insertion loss, and efficient electro-optic modulation. However, due to its crystal symmetry, silicon lacks a linear electro-optic effect (Pockels effect). Therefore, plasma-dispersion in *p-n* junctions is employed. However, this approach leads to a detrimental coupling of phase modulation and amplitude modulation, and it is usually less energy efficient than modulation via the Pockels effect.

In this work we go beyond the limits of CMOS-processing for reducing the energy consumption of silicon PICs. A novel fabrication process for silicon nanowire waveguides is developed. The waveguides have ultra-smooth sidewalls and low propagation loss. Furthermore, the hybridization of CMOS-processed silicon PICs with electro-optic organic materials is investigated and results in highly energy efficient hybrid modulators. The thesis is structured as follows:

Chapter 1 gives an overview of the state of the art in PIC fabrication, fiber-chip coupling, and high-speed modulators in silicon.

Chapter 2 summarizes the theoretical and technological background. The chapter covers second-order nonlinear effects in organic materials, an overview on common waveguide geometries, the basic methods for coupling silicon PICs to optical fibers, and the operation principle of a silicon-organic hybrid modulator.

Chapter 3 presents a novel fabrication method that utilizes anisotropic wet etching instead of conventional dry etching. The strong etching anisotropy of an aqueous potassium hydroxide solution is used to fabricate ultra-smooth waveguides with trapezoidal cross-section. In addition, highly efficient grating couplers with trapezoidal etch grooves are studied and demonstrated.

Chapter 4 demonstrates a fiber-chip coupling scheme that allows for an in-plane assembly of optical fiber and silicon PIC. Total internal reflection at the facet of an angle-polished fiber is exploited for coupling light perpendicularly to the fiber axis to a silicon grating coupler on the chip surface. The coupling efficiency and the misalignment tolerances of this in-plane coupling scheme are investigated.

Chapter 5 focusses on silicon-organic hybrid (SOH) Mach-Zehnder modulators (MZM) for high-speed and energy-efficient signalling. The chapter covers the operation principle of an SOH push-pull MZM, the design of low-loss auxiliary building blocks, the applicability and the poling of different organic materials, the design of RF transmission lines, and the electro-optic frequency response of the modulator. The fabricated SOH modulators are tested in system experiments using different modulation formats, such as on-off keying (OOK) and amplitude-shift-keying (ASK) at data rates up to 84 Gbit/s. At a data rate of 12.5 Gbit/s we demonstrate electro-optic modulation with an energy-efficiency down to 0.7 fJ/bit.

Chapter 6 summarizes the main results of this thesis and gives an outlook on further work.

Achievements of the Present Work

In this thesis, methods for reducing the optical and electrical power consumption of silicon photonic integrated circuits have been investigated. We employed a novel fabrication concept for strongly guiding silicon nanowires, investigated an efficient in-plane fiber-chip coupler, and demonstrated high-speed, low-power electro-optic modulators in silicon-organic hybrid (SOH) technology.

In the following, we give a concise overview of the main achievements.

Waveguide Fabrication: Fabrication scheme for silicon nanowire waveguides using anisotropic wet etching

Grating Couplers: First-time demonstration of grating couplers fabricated by anisotropic wet etching. Predicted fiber-chip coupling losses amount to 0.3 dB when using a non-uniform grating in combination with a metallic reflector beneath the buried oxide layer.

Staggered Mask Scheme: Lithographic masks with discrete staggered elements enable wet chemical fabrication of nanowire waveguide by low-resolution contact lithography.

Angled Fibers: Development and analysis of a horizontal fiber-chip coupler that utilizes specially polished angled fibers in combination with on-chip grating couplers.

Coplanar Transmission Line: Design and characterization of coplanar transmission lines for SOH phase modulators, Mach-Zehnder modulators, and IQ modulators. Analysis of device and substrate contribution to RF loss.

Passive Optical Components: Design and characterization of low-loss passive optical elements like multimode interference couplers and mode converters. Using a special, logarithmically tapered strip-to-slot waveguide mode converter, a converter loss of 0.02 dB is demonstrated [C15, J4].

Poling of Organic Materials: Development of a poling scheme for SOH push-pull Mach-Zehnder modulators (MZM) [J5, J7]. Hybridization of novel shape- and structurally engineered chromophore molecules with silicon slot waveguides [C28, C30]. This approach resulted in the highest in-device electro-optic coefficient of 230 pm/V [C30] reported so far.

High-Speed SOH Modulator: We fabricate the first high-speed SOH MZM and demonstrate operation at 40 Gbit/s using on-off keying [J5]. Furthermore, we demonstrate that a single SOH MZM suffices for multilevel signalling with a record data rate of 84 Gbit/s [C21, J5].

Energy-efficient SOH Modulator: Demonstration of an SOH MZM [C28] with an energy consumption below 1 fJ / bit. This is the lowest reported energy consumption of a non-resonant silicon modulator.

Chip-Design: Contribution to design and layout of photonic integrated circuits (MMI, mode converters, electrical isolation of optical WG, RF transmission lines, grating couplers) for fabrication in CMOS foundries.

1 Introduction

Data traffic in world-wide communication networks continues to grow exponentially and represents one of the biggest challenges of contemporary information and communication technology. Optical interconnects based on photonic integrated circuits (PIC) are considered the most viable solution to cope with these challenges by increasing transmission capacity while keeping the associated energy consumption at an acceptable level. The integrated components need to feature functionality and performance comparable or even superior to those of discrete optical components. In addition, footprint, energy consumption and cost of integrated devices have to be significantly smaller than those of their discrete counterparts.

Silicon-on-insulator (SOI) is expected a promising platform to fulfil these demands. The compatibility of silicon photonics to CMOS-based large-scale fabrication lends itself to the realization of cost-effective circuits, possibly co-integrated with CMOS-driver electronics. The high refractive index contrast between silicon waveguide core ($n = 3.5$) and dielectric cladding ($n \approx 1.5$) results in strongly guided modes and in a high integration density. Recent research on silicon-photonics has been focussed on increasing the performance of silicon PIC and on *reducing the associated optical and electrical energy consumption.*

1.1 SOI Waveguide Fabrication: State of the Art

Fabrication of silicon photonics PIC widely relies on mature CMOS processes exploiting 193 nm deep UV lithography in combination with inductively coupled reactive ion plasma etching [3]. The existing CMOS infrastructure that was originally developed for the fabrication of electronic circuitry gave rise to the photonic foundry services that offer the unique possibility of "fabless" PIC fabrication to companies and academic institutions that do not possess the required technological infrastructure and fabrication expertise. Examples of foundries are ePIXfab, using the technology of IMEC in Leuven (Belgium) and LETI in Grenoble (France), as well as OpSIS, using the CMOS facilities of the Institute of Microelectronics (IME) in Singapore. As an alternative to deep-UV lithography fabrication, companies like AMO GmbH in Aachen (Germany) and IMS-chips in Stuttgart (Germany) offer foundry services based on electron-beam lithography.

For most applications the insertion loss of a PIC has to be sufficiently low. The insertion loss has three major contributions: Fiber-to-chip coupling loss, mode-mismatching loss and scattering loss at rough surfaces, and material absorption. Propagation loss of silicon strip waveguides is dominated by scattering loss due to sidewall roughness of the waveguides. Typical propagation losses of foundry-fabricated strip waveguides are in the order of 2 dB/cm, while recent technological improvements achieved with 193 nm deep-UV immersion lithography result in losses as low as 0.7 dB/cm [4]. For bridging large distances on the chip, shallow-etched rib waveguides are particularly suited. They represent a weakly guided waveguide structure and have low propagation loss of 0.3 dB/cm [3].

Alternative approaches to achieve low propagation loss comprise excimer laser reformation techniques to smoothen the sidewall roughness in a post-fabrication step, micro-scale ridge waveguides, so-called "etchless" waveguide fabrication by aggressive thermal oxidation, and anisotropic wet etching.

Excimer laser induced reflow of the silicon waveguides has been demonstrated to result in an rms roughness reduction from 14 nm to 0.24 nm [5]. From this roughness a propagation loss of 0.03 dB/cm is predicted by coupled mode theory [6]. However, this surface reformation is not scalable for mass production, and it results in a round waveguide cross-section that is not compatible to every component of the integrated circuit.

Micro-scale ridge waveguides built from up to 11 μm thick SOI layers are another example for a weakly guided waveguide structure. This waveguide geometry has a propagation loss of 0.1 dB/cm [7] and a low loss of 0.2 dB [7] when coupled to a standard singlemode fiber (SSMF). However, the large waveguide cross-section comes at the prize of a small integration density and large bend radii, resulting in a large footprint of the PIC.

By using *"etchless" waveguides*, fabricated by inhomogeneous oxidation of a 500 nm thick SOI layer, a low propagation loss of 0.3 dB/cm has been demonstrated [8]. The resulting waveguide geometry is 1 μm wide and has a height of only 80 nm. The measured rms surface roughness is 0.3 nm. The mode of this waveguide is only weakly confined, resulting once more in a low integration density. Furthermore, the fabrication process challenges the implementation of additional elements like gratings, modulators, and detectors.

Anisotropic wet etching based on aqueous potassium hydroxide solutions is mainly known from the fabrication of micro-mechanic elements like microfluidic channels, alignment V-grooves, etc. The high selectivity with respect to crystal-planes during etching could in principle be used to fabricate strongly guiding silicon waveguides with atomically flat sidewalls. So far, demonstrations of waveguide fabrication with anisotropic wet etching was limited to micro-scale ridge waveguides with large cross-section [9]. The applicability of this approach for the fabrication of nanowire waveguides is *studied in this thesis.*

1.2 Fiber-Chip Coupling: State of the Art

Efficient coupling between an integrated waveguide and an external glass fiber is a key requirement for any photonic integrated circuit platform. In silicon photonics mainly two approaches have been used for efficient fiber-chip coupling – inverted taper spot-size converters and grating couplers.

Inverted tapers denote a singlemode waveguide transition from a strongly guiding narrow cross-section to an even narrower and therefore much weaker guiding structure. The approach features efficient in-plane fiber-chip coupling and a large optical bandwidth. On the other hand, inverted taper couplers require cleaved fiber facets, which makes wafer-scale testing of devices challenging. An elegant SSMF coupling solution with passive fiber alignment has been demonstrated by etching an alignment V-groove into the silicon substrate of an SOI sample [10]. The remaining mode size mismatch between the inverted taper in the chip and the SSMF results in moderate coupling losses of 3.5 dB [10]. High-coupling efficiencies usually require the additional use of tapered and/or lensed fibers that result in additional product cost. However, losses down to 1 dB [11] have been reported. Recent developments in three dimensional direct laser writing based on two-photon polymerization offer new possibilities for fiber-chip coupling: An on-chip inverted taper connects to a SSMF with a 3D polymer waveguide taper that widens at the fiber endface. For this technology even smaller low coupling losses are to be expected. This directly-written polymer waveguide is a special form of a so-called photonic wire bond [12], where chip-to-chip interconnects with coupling losses of 1.6 dB could be demonstrated.

Grating couplers, in contrast to inverted tapers, enable wafer-scale testing of devices by close-to-vertical fiber-chip coupling. No tapered fibers are required, and experimental chip-to-SSMF coupling efficiencies of 0.6 dB have been demonstrated [13]. Furthermore, the coupling tolerances between grating and fiber allow for a misalignment of the fiber in the order of 2 μm. Disadvantages of this approach are the requirement of nearly vertical chip-coupling, which is impractical for packaging, and the limited optical bandwidth of a grating. A workaround to the vertical assembly of fiber and chip was developed and studied *in this thesis* by making use of total internal reflection at the polished angled facet of a SSMF. This concept was independently developed at the Tyndall National Institute and published in 2012 [14].

1.3 Silicon Electro-Optic Modulators: State of the Art

For designing an energy-efficient electro-optic modulator in silicon, the restrictions imposed by the choice of this material have to be observed. While commercial amplitude modulators made from materials like $LiNbO_3$, InP or GaAs rely on the ultra-fast Pockels effect – a second-order nonlinear optical effect – this effect cannot be exploited in crystalline silicon. The reason for this is that second-order nonlinearities are absent in bulk silicon due to the inversion symmetry of the crystal lattice. Consequently, research activities that focus on electro-optic modulation can be split in two major groups. The first group realizes electro-optic modulation in silicon by exploiting plasma dispersion, while the second group focusses on hybridizing silicon with a second material system. Another method applies strain to the silicon crystal lattice by growing silicon nitride on silicon, which leads to a certain degree of second-order nonlinearity [15], [16]. However, the resulting linear electro-optic effect is still too small for a practical application.

Plasma-dispersion

Most silicon modulators are based on the *plasma-dispersion-effect*. Here, charge carrier injection or depletion in a diode structure changes the amount of free carriers in a *p-n* junction which forms part of a silicon waveguide. Both the absorption and the refractive index of the silicon waveguide core are influenced [17]. This leads to various trade-offs when realizing fast and energy-efficient devices with small footprint: Carrier injection in forward-biased *p-i-n* structures enables compact modulators with voltage-length products as low as $U_\pi L = 0.36\,\text{Vmm}$ [18], but the relatively large free-carrier lifetime limits the operation frequency to a few GHz [19]. Moreover, the energy efficiency of these devices is intrinsically limited to several pJ/bit due to the permanent injection current flowing through the diode section. In contrast, reverse-biased *p-n* junctions with carrier depletion have negligible bias currents and support symbol rates of up to 50 GBd [20]. As a downside, typical voltage-length products are larger than 10 Vmm and thus much worse than those of injection-type devices. Despite of the rather large footprint of carrier depletion devices, their basic feasibility for the realization of integrated IQ-modulators has been shown by QPSK and 16QAM modulation at a symbol rate of 28 GBd [21], [22]. An energy consumption of 200 fJ/bit has been achieved in a 5 mm-long depletion-type Mach-Zehnder modulator (MZM) [23]. Modulation energies and device footprint can be significantly reduced by using resonant structures. For resonant silicon-based modulators, the lowest energy consumption reported to date amounts to 3 fJ/bit and has been achieved with a microdisc device [24] operated at a data rate of 12.5 Gbit/s. However, the use of resonant devices comes at the price of limited optical bandwidth and low fabrication yield. Furthermore, the resonance wavelength is often subject to strong temperature-induced drifts.

Hybrid Integration

An alternative to the plasma-dispersion effect is the hybrid integration of silicon with a second material system. By exploiting the optical properties of the addtitional material new functionalities like second-order nonlinearity can be added. Examples are the hybrid integration of III-V compounds [25] or organic materials [26] for realizing energy efficient modulators with small voltage-length product. In this thesis we investigate the hybridization of silicon slot waveguides [27] with organic materials that feature a strong electro-optic effect [28]. Basically, the silicon-organic hybrid (SOH) approach can simultaneously meet the modulator requirements of low drive voltage, high optical and electrical bandwidth, compact size and low energy consumption [26], [29]. While the basic operation principle of modulation in SOH slot waveguides has been demonstrated in various publications [30]–[35] these devices where still far from exploiting the full potential of the SOH approach: The reported modulators featured small in-device electro-optic coefficients ($r_{33} < 60$ pm/V), resulting in either a high operating drive voltage or in a large device size comparable to or worse than those of carrier-depletion plasma-effect modulators. Also, the electro-optic bandwidth of most devices was in the order of few GHz [36]. In this thesis, the issues of small electro-optic responsivity and of small modulation speeds are addressed by a proper design of optical and electrical device characteristics, by investigation of novel organic material concepts for SOH integration and by the development of an efficient poling scheme. The developed MZM have an electro-optic bandwidth larger than 20 GHz and enable a data rate of 84 Gbit/s in a single polarization [C21, J5], the so far highest data rate reported for a silicon MZM. In addition, the fabricated modulators feature record-high in-device electro-optic coefficients of up to $r_{33} = 230$ pm/V [C28, C30], resulting in a voltage-length product of $U_{\pi}L = 0.5\,\text{Vmm}$ and in an energy consumption as low as 1.6 fJ/bit [C28], the currently smallest value for a MZM in any material system.

2 Theoretical and Technological Background

This chapter covers the theoretical and technological background of this thesis and is structured as follows.

Section 2.1 gives a brief introduction into linear and nonlinear polarization and into electro-optic activity in organic materials. A more detailed introduction to nonlinear effects in bulk materials is given by the textbook of R. W. Boyd [37]. For a more immersed understanding of nonlinear phenomena in organic materials the comprehensive paper of J. Wolff and R. Wortmann [38] is recommended. In Section 2.2 an overview on waveguiding structures on the Silicon-on-Insulator (SOI) platform is given. For a general introduction to waveguiding in integrated photonic circuits the textbook of J-M. Liu [39] is recommended. Section 2.3 covers the concepts of spot-size converting structures for coupling silicon waveguides to fibers. Finally, the concept of a Silicon-Organic hybrid (SOH) Mach-Zehnder Modulator (MZM) is explained in Section 2.4, along with an analytic model for predicting its electro-optic frequency response. In the following, we use the electrical engineering time dependency $\mathrm{e}^{+\mathrm{j}\omega t}$.

2.1 Linear and Nonlinear Electric Polarization

The electric polarization is the response of a medium to an external electric field. The time-dependent electric polarization $\vec{P}$ as a function of the electric field $\vec{E}$ can be described by a Volterra series, accounting for possible memory effects of the nonlinear medium that are described by causal influence functions $\underline{\chi}^{(n)}(t', t'', ..., t^{(n)})$ (tensors of rank n+1),

$$\begin{aligned}
\vec{P}(\vec{r},t) = {} & \varepsilon_0 \int_0^{+\infty} \underline{\chi}^{(1)}(\vec{r},t-t')\vec{E}(\vec{r},t')\,\mathrm{d}t' \\
& +\varepsilon_0 \int_0^{+\infty}\int_0^{+\infty} \underline{\chi}^{(2)}(\vec{r},t-t',t-t''):\vec{E}(\vec{r},t')\vec{E}(\vec{r},t'')\,\mathrm{d}t'\,\mathrm{d}t'' \\
& +\varepsilon_0 \int_0^{+\infty}\int_0^{+\infty}\int_0^{+\infty} \underline{\chi}^{(3)}(\vec{r},t-t',t-t'',t-t'''):\vec{E}(\vec{r},t')\vec{E}(\vec{r},t'')\vec{E}(\vec{r},t''')\,\mathrm{d}t'\,\mathrm{d}t''\,\mathrm{d}t''' \\
& +\ldots ,
\end{aligned} \tag{2.1}$$

The vacuum permittivity is $\varepsilon_0 = 8.85419 \times 10^{-12}\,\mathrm{As/(Vm)}$. The influence functions define the temporal memory of the material and are causal,

$\underline{\chi}^{(n)}(\vec{r},t',t'',...,t^{(n)}) = \mathbf{\underline{0}}$ for $t',t'',t'''... < 0$, which is also expressed by the lower integral boundaries in Eq. (2.1).

2.1.1 Linear Polarization

The first-order influence function $\underline{\chi}^{(1)}$ is usually several orders of magnitude larger than the influence functions of higher orders. Therefore we split Eq. (2.1) into a linear term $\vec{P}^{(1)}$ and into a nonlinear term $\vec{P}^{\mathrm{NL}}$

$$\vec{P}(\vec{r},t) = \vec{P}^{(1)}(\vec{r},t) + \vec{P}^{\mathrm{NL}}(\vec{r},t)\ . \tag{2.2}$$

The linear polarization $\vec{P}^{(1)}$ takes the form

$$\vec{P}^{(1)}(\vec{r},t) = \varepsilon_0 \int_0^{+\infty} \underline{\chi}^{(1)}(\vec{r},t-t')\vec{E}(\vec{r},t')\mathrm{d}t'\ . \tag{2.3}$$

In the Fourier transformation (A.1), Eq. (2.3) reads

$$\tilde{\vec{P}}^{(1)}(\vec{r},\omega) = \varepsilon_0 \underline{\tilde{\chi}}^{(1)}(\vec{r},\omega)\tilde{\vec{E}}(\vec{r},\omega), \tag{2.4}$$

with

$$\underline{\tilde{\chi}}^{(1)}(\vec{r},\omega) = \int_0^{+\infty} \underline{\chi}^{(1)}(\vec{r},t)e^{-\mathrm{j}\omega t}\,\mathrm{d}t\ . \tag{2.5}$$

The complex analytic function $\underline{\tilde{\chi}}^{(1)}(\omega) = \underline{\tilde{\chi}}_{\mathrm{r}}^{(1)}(\omega) + \mathrm{j}\underline{\tilde{\chi}}_{\mathrm{i}}^{(1)}(\omega)$ is called optical susceptibility of first order and it is defined by the complex permittivity $\underline{\varepsilon}_{\mathrm{r}}(\omega) - \mathrm{j}\underline{\varepsilon}_{\mathrm{i}}(\omega) = \mathbf{\underline{1}} + \underline{\tilde{\chi}}_{\mathrm{r}}^{(1)}(\omega) + \mathrm{j}\underline{\tilde{\chi}}_{\mathrm{i}}^{(1)}(\omega)$. Due to the causality of the influence function $\underline{\chi}^{(1)}(\vec{r},t')$ real and imaginary part of the first-order optical susceptibility $\underline{\tilde{\chi}}^{(1)}(\vec{r},\omega)$ are connected by the Hilbert transform (Kramers-Kronig relations, Eq. (A.3)). As a result, material gain or absorption is connected with material dispersion. Consequently, the only medium that has a constant refractive index for all frequencies is vacuum, a "medium" without memory. All other materials have a memory and therefore feature a non-vanishing frequency dependent real and imaginary part of the optical susceptibility. However, the susceptibility of a medium can be nearly constant $\underline{\tilde{\chi}}^{(1)}(\omega) \approx \underline{\tilde{\chi}}^{(1)}(\omega_0) = \mathrm{const}$ in a certain frequency range where the material absorption is negligible. With this assumption we transform Eq. (2.4) back into time domain. The result is a linear relation between polarization and electric field in time domain,

$$\vec{P}^{(1)}(\vec{r},t) = \varepsilon_0 \underline{\tilde{\chi}}^{(1)}(\vec{r},\omega_0)\vec{E}(\vec{r},t). \tag{2.6}$$

Note that Eq. (2.6) actually describes a medium without memory, and the polarization follows the electric field instantaneously.

Using this result and neglecting $\vec{P}^{\mathrm{NL}}$, the displacement field $\vec{D}$ can be expressed as

$$\vec{D}(\vec{r},t) = \varepsilon_0 \vec{E}(\vec{r},t) + \vec{P}(\vec{r},t) = \varepsilon_0 \left(\underline{\mathbf{1}} + \underline{\tilde{\chi}}^{(1)}(\vec{r},\omega_0) \right) \vec{E}(\vec{r},t) = \varepsilon_0 \underline{\varepsilon}_r \vec{E}(\vec{r},t), \quad (2.7)$$

where $\underline{\varepsilon}_r$ is the dielectric permittivity tensor. For an isotropic medium permittivity and susceptibility can be treated as scalar quantities and the complex refractive index $n - \mathrm{j}n_\mathrm{i}$ of the medium can be expressed by

$$\begin{aligned} &\varepsilon_\mathrm{r} - \mathrm{j}\varepsilon_\mathrm{i} = (n - \mathrm{j}n_\mathrm{i})^2 = n^2 - 2\,\mathrm{j}nn_\mathrm{i} - n_\mathrm{i}^2 \\ &\varepsilon_\mathrm{r} = n^2 - n_\mathrm{i}^2 \approx n^2 \quad \rightarrow \quad n = \sqrt{\varepsilon_\mathrm{r}}, \text{ if } n_\mathrm{i} \approx 0 \\ &\varepsilon_\mathrm{i} = 2nn_\mathrm{i} \quad \rightarrow \quad n_\mathrm{i} = \varepsilon_\mathrm{i} / (2n) \end{aligned} \quad (2.8)$$

2.1.2 Nonlinear Polarization

Referring to Eq. (2.1) and Eq. (2.2) the nonlinear polarization takes the form

$$\begin{aligned} \vec{P}^{\mathrm{NL}}(\vec{r},t) &= \sum_{n=2}^{\infty} \vec{P}^{(n)} \\ &= \varepsilon_0 \int_0^{+\infty}\int_0^{+\infty} \underline{\chi}^{(2)}(\vec{r},t-t',t-t'') : \vec{E}(\vec{r},t')\vec{E}(\vec{r},t'')\,\mathrm{d}t'\,\mathrm{d}t'' \\ &+ \varepsilon_0 \int_0^{+\infty}\int_0^{+\infty}\int_0^{+\infty} \underline{\chi}^{(3)}(\vec{r},t-t',t-t'',t-t''') \vdots \vec{E}(\vec{r},t')\vec{E}(\vec{r},t'')\vec{E}(\vec{r},t''')\,\mathrm{d}t'\,\mathrm{d}t''\,\mathrm{d}t''' \\ &+ \ldots \end{aligned} \quad (2.9)$$

We restrict ourselves to second-order nonlinear effects $\left(\underline{\chi}^{(n)} = \underline{\mathbf{0}}\ \forall\ n > 2\right)$. Assuming that the frequency spectrum of the input electric field is composed of discrete angular frequencies $\omega_1, \omega_2 \ldots \omega_n$, the Fourier transform of Eq. (2.9) can be written as ([37], p. 19, Eq. (1.3.12))

$$\tilde{P}_i^{(2)}(\omega_n + \omega_m) = \varepsilon_0 \sum_{jk} \sum_{(nm)} \tilde{\chi}_{ijk}^{(2)}(\omega_n + \omega_m; \omega_1, \omega_2) \tilde{E}_j(\omega_n) \tilde{E}_k(\omega_m). \quad (2.10)$$

The spatial dependence of the quantities above is omitted for the sake of readability. $\tilde{P}^{(2)}$ is the second-order polarization, and $\underline{\tilde{\chi}}^{(2)}$ is the nonlinear susceptibility tensor of second order. The indices $i, j, k = \{1,2,3\}$ denote the three orthogonal coordinate axes. The notation (nm) indicates that the summation over the indices n and m includes only those sets of frequencies ω_n, ω_m that result in a frequency $\omega_n + \omega_m$. Next, we will investigate a special second-order nonlinear effect, the so-called Pockels effect.

2.1.3 Pockels Effect

The linear electro-optic effect, also called Pockels effect, denotes the change of refractive index of a material due to the presence of a slowly varying electric field. The change of the refractive index is a result of the second-order nonlinear interaction of a low-frequency electrical field with $\omega_\mathrm{m} \approx 0$ and a high-frequency optical field with ω_c in a material with non-zero $\underline{\tilde{\chi}}^{(2)}$-tensor. In this case the second-order nonlinear polarization, Eq. (2.10), becomes

$$\tilde{P}_i^{(2)}(\omega_\mathrm{c} + \omega_\mathrm{m} \approx \omega_\mathrm{c}) = 2\varepsilon_0 \sum_{jk} \tilde{\chi}_{ijk}^{(2)}(\omega_\mathrm{c} + \omega_\mathrm{m}; \omega_\mathrm{c}, \omega_\mathrm{m}) \tilde{E}_j(\omega_\mathrm{c}) \tilde{E}_k(\omega_\mathrm{m}). \tag{2.11}$$

Here, the factor of 2 is a result of the permutation symmetry of the input fields, which results in a permutation symmetry of the susceptibility tensor.

$$\tilde{\chi}_{ijk}^{(2)}(\omega_\mathrm{c} + \omega_\mathrm{m}; \omega_\mathrm{c}, \omega_\mathrm{m}) = \tilde{\chi}_{ikj}^{(2)}(\omega_\mathrm{c} + \omega_\mathrm{m}; \omega_\mathrm{m}, \omega_\mathrm{c}). \tag{2.12}$$

In the following we use the Einstein notation for sums. Using Eq. (2.11) the displacement field spectrum $\tilde{D}_i(\omega)$ can be expressed by

$$\begin{aligned} \tilde{D}_i(\omega_\mathrm{c}) &= \varepsilon_0 \left(\delta_{ij} + \tilde{\chi}_{ij}^{(1)}(\omega_\mathrm{c}) \right) \tilde{E}_j(\omega_\mathrm{c}) \\ &\quad + 2\varepsilon_0 \tilde{\chi}_{ijk}^{(2)}(\omega_\mathrm{c} + \omega_\mathrm{m}; \omega_\mathrm{c}, \omega_\mathrm{m}) \tilde{E}_j(\omega_\mathrm{c}) \tilde{E}_k(\omega_\mathrm{m}). \end{aligned} \tag{2.13}$$

We define a low-frequency field-dependent permittivity ε_{ij}

$$\begin{aligned} \varepsilon_{ij} &= \varepsilon_0 \left(\delta_{ij} + \tilde{\chi}_{ij}^{(1)} + 2\tilde{\chi}_{ijk}^{(2)} \tilde{E}_k(\omega_m) \right) \\ &= \varepsilon_0 \left(\varepsilon_{\mathrm{r},ij} \qquad + \Delta\varepsilon_{\mathrm{r},ij} \right). \end{aligned} \tag{2.14}$$

The nonlinear change of the permittivity $\Delta\underline{\varepsilon}_\mathrm{r}$ is given by

$$\Delta\varepsilon_{\mathrm{r},ij} = 2\tilde{\chi}_{ijk}^{(2)} \tilde{E}_k(\omega_\mathrm{m}), \tag{2.15}$$

and is therefore proportional to the low-frequency electric field $\tilde{E}_k(\omega_\mathrm{m})$.

For the special case where both input fields are polarized along axis 3 and if the dominant tensor component is $\tilde{\chi}_{333}^{(2)}$, the change of the permittivity is given by

$$\Delta\varepsilon_\mathrm{r} = 2\tilde{\chi}_{333}^{(2)} \tilde{E}_3(\omega_\mathrm{m}). \tag{2.16}$$

The nonlinear perturbation Δn of the refractive index $n + \Delta n$ can be approximated by

$$\varepsilon_\mathrm{r} + \Delta\varepsilon_\mathrm{r} = (n + \Delta n)^2 = n^2 + 2n\,\Delta n + \Delta n^2 \approx n^2 + 2n\Delta n, \tag{2.17}$$

Substituting Eq. (2.16) in Eq. (2.17) we get an expression for the Pockels effect induced change of refractive index,

$$\Delta n = \frac{\Delta \varepsilon_{\mathrm{r}}}{2n} = \frac{\tilde{\chi}_{333}^{(2)}}{n} \tilde{E}_3(\omega_{\mathrm{m}}) . \tag{2.18}$$

We find that the change of the refractive index is indeed proportional to the slowly varying electric field.

For historical reasons the use of the electro-optic coefficient r_{ijk} is usually preferred over the second-order susceptibility $\tilde{\chi}_{ijk}^{(2)}$ (see Ref. [37], Chapter 11). According to Ref. [37], Eq. (11.3.7a), and Ref. [40], Eq. (3.15) the change of refractive index due to an external electric field E_k is given by

$$\Delta n_{ij} = -\frac{1}{2} n^3 r_{ijk} E_k(\omega_{\mathrm{m}} \approx 0) . \tag{2.19}$$

By comparing Eq. (2.19) and Eq. (2.18) we find that the electro-optic coefficient and the second-order susceptibility are linked by the following relation

$$r_{ijk} = -\frac{2\chi_{ijk}^{(2)}}{n^4} . \tag{2.20}$$

When considering permutation symmetry of the input fields, the independent components of the tensors $\underline{\tilde{\chi}}^{(2)}$ and $\underline{r}$ are reduced from 27 to 18. The tensor components can then be described in a contracted notation $\tilde{\chi}_{ih}^{(2)}$, r_{ih} by only two indices, where the index h is defined by

$$h = \begin{cases} 1 \text{ for } jk = 11, \\ 2 \text{ for } jk = 22, \\ 3 \text{ for } jk = 33, \\ 4 \text{ for } jk = 23 \text{ or } 32, \\ 5 \text{ for } jk = 13 \text{ or } 31, \\ 6 \text{ for } jk = 12 \text{ or } 21. \end{cases} \tag{2.21}$$

Next, we will explain how the macroscopic electro-optic coefficient of an organic material is linked to the optical properties of the ensemble of individual nonlinear molecular dipoles.

2.1.4 Second-Order Nonlinear Organic Materials

The following subsection is based on [38], where more detailed information on organic electro-optic materials can be found. Here, only a brief summary of the microscopic nonlinear properties of organic materials is given. In most cases electro-optic organic molecules are chromophores (Greek for "pigment"). As the name indicates, these molecules have a very colourful appearance that has its origin in a strong material absorption in the region of visible light. Essential for a nonlinear chromophore is the existence of a donor group and of an acceptor group that are linked by an extended π-conjugated bridge. This π-conjugated bridge allows the electronic coupling of donor and acceptor group and thus a transfer of an electron from donor to acceptor group. An exemplary molecule is depicted in Fig. 2.1(a) and (b). Here, tetraene is used as a π-conjugated bridge. We differentiate between the so-called "polyene" state of the chromophore, where the molecular dipole moment is zero and the "betaine" state, where an electron is transferred from donor to acceptor, inducing a molecular dipole moment. The real distribution of the electron wave function usually neither corresponds to the polyene state nor to the betaine state, but is rather a superposition of the two states.

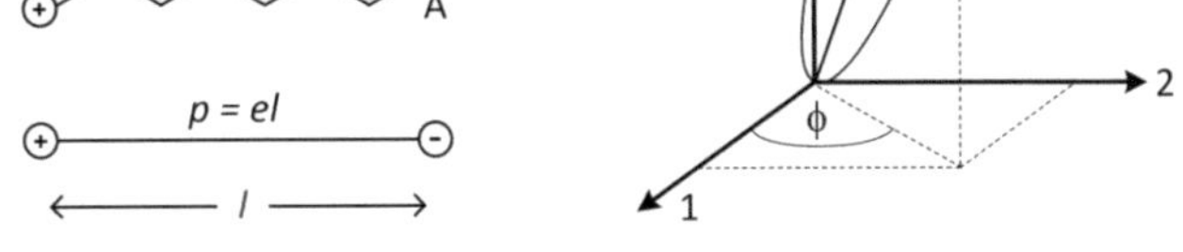

Fig. 2.1 Eigenstates of a nonlinear chromophore molecule. (a) "Polyene state" – the molecule is neutral, no molecular dipole moment exists. (b) "Betaine" state – An electron of charge e is transferred from donor D to acceptor A. The betaine state has a molecular dipole moment $p = el$. (c) Orientation of the molecular coordinate system (x,y,z) with respect to the global coordinate system (1,2,3). The ellipsoid represents a nonlinear molecule, and the z-axis is the axis of the dipole moment. An external poling field along the global axis 3 forces the molecular dipole moment to align with the global axis 3. The angle θ denotes the residual angle between axis 3 and axis z. Figure redrawn after [41], Fig. 3.5.

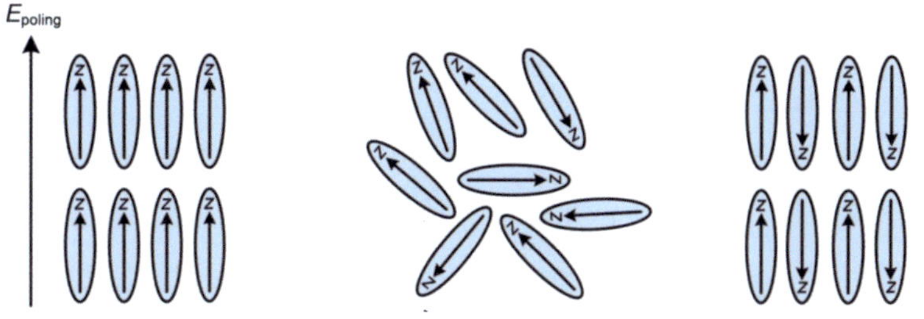

Fig. 2.2 Molecular ordering. (a) Acentric ordering due to chromophore-poling field interaction. (b) Isotropic ordering caused by thermal randomization. (c) Centric ordering due to chromophore-chromophore electrostatic interaction. Figure adapted from [159].

Similar to the macroscopic description of nonlinear effects in Eq. (2.10) the microscopic nonlinear dipole moment of a molecule can be described in terms of a power series,

$$p_i = \mu_i + \alpha_{ij} E_j + \beta_{ijk} E_j E_k + \gamma_{ijkl} E_j E_k E_l + \ldots, \tag{2.22}$$

where $\vec{p}$ is called molecular polarizability, $\vec{\mu}$ is the electric dipole moment, $\underline{\alpha}$ is the first-order polarizability, $\underline{\beta}$ is the second-order polarizability (or first hyperpolarizability) and $\underline{\gamma}$ is the third-order polarizability (or second hyperpolarizability). Here, the indices $i, j, k = \{x, y, z\}$ denote the three orthogonal coordinate axes of the molecule, as depicted in Fig. 2.1(c). The second-order nonlinear effects in organic materials thus have their origin in the second-order polarizability $\underline{\beta}$. Elongated chromophores with strong donor and acceptor groups at the terminal positions usually possess only one significant component β_{zzz} along the dipole axis. The dipole density determines the macroscopic electric polarization, therefore β_{zzz} is linked to the second-order susceptibility by

$$\tilde{\chi}_{333}^{(2)} = Ng \left\langle \cos^3 \theta \right\rangle \beta_{zzz}. \tag{2.23}$$

The quantity N is the number density of the nonlinear molecular dipoles in the material and g is the Lorentz-Onsager local field factor that corrects for partial screening of the external electric field. The average acentric order parameter $\left\langle \cos^3 \theta \right\rangle$ of the molecule ensemble denotes the average orientation of the molecules with respect to an external field, and θ is the angle between input field polarization (aligned to axis 3) and molecular dipole axis z, as depicted in Fig. 2.1(c). The acentric order parameter takes account of the fact that a macroscopic EO activity can only exist for anisotropically and acentrically ordered molecule ensembles, see Fig. 2.2. For an isotropic or centric order of the molecules the acentric order parameter is zero, and thus no macroscopic nonlinearity of second order exists.

Unfortunately, the thermodynamically stable state of the ensemble is the centric order, where the molecular dipole moments cancel in a so-called "head-to-tail" orientation, Fig. 2.2(c). A global orientation of the dipoles (acentric order), however, can be achieved in a so-called poling process, Fig. 2.2(a). For poling, the material is heated to its glass transition temperature while a strong static electric poling field E_{poling} is applied. After the molecular dipoles have oriented according to the external field, the material is cooled down while maintaining the poling field. This way the acentric order remains. Combining Eq. (2.20) and Eq. (2.23) the macroscopic linear EO coefficient r_{33} is given by

$$r_{33} = \frac{-2\chi_{333}^{(2)}}{n_0^4} = -2N\beta_{zzz}\left\langle \cos^3\theta \right\rangle \frac{g}{n_0^4}. \tag{2.24}$$

Therefore the macroscopic EO coefficient of an organic material depends on the microscopic EO second-order polarizability β_{zzz} of a single chromophore molecule, on the dipole number density N, and on the orientation of the chromophore ensemble with respect to an external field.

2.2 Silicon-on-Insulator (SOI) Waveguides

The SOI platform was originally developed for integrated electronic CMOS circuits. The wafer consists of a crystalline silicon layer with a thickness of a few tens of nanometers, which is separated from the bulk silicon substrate by a thin layer of silicon dioxide (buried oxide, BOX), typically few 10 nm to 150 nm [42] in thickness. These SOI substrates have to be modified for building optical waveguide structures in silicon. Optical waveguides require a high refractive index waveguide core and a low refractive index cladding material. In the case of integrated silicon waveguides the cladding consists of an upper part that is deposited after definition of the waveguide core and of a lower cladding — the buried oxide. For strongly guiding silicon waveguides, the thickness of the buried oxide layer needs to be at least 1 µm to avoid mode leakage into the silicon substrate. For the upper cladding usually silicon dioxide or an organic material is used [3], [26]. Typical SOI substrates for photonic integrated circuits (PIC) have a BOX thickness of 2 µm or 3 µm and have a device layer thickness of 220 nm [42] for devices that are intended to guide

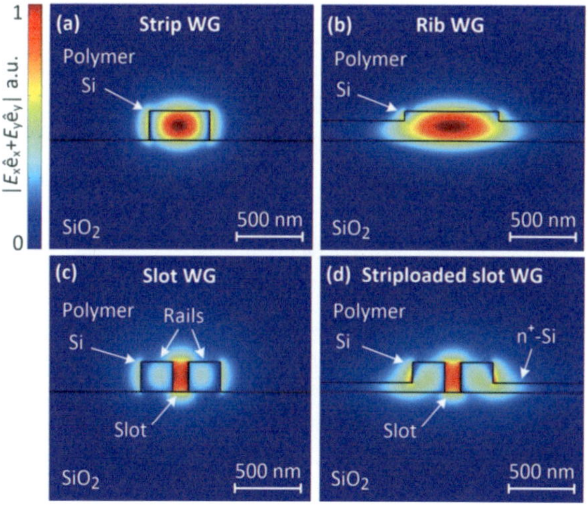

Fig. 2.3 Optical waveguide geometries on the SOI platform and simulated quasi-TE polarized electric mode field. (a) Strip WG. Dimensions: 220 nm × 450 nm. (b) Rib WG. The WG width is 700 nm and the etch depth is 70 nm. (c) A slot WG consists of two silicon rails separated by a narrow slot. The mode is strongly confined to the slot region. Here, the rail width is 240 nm and the slot width is 120 nm. (d) A striploaded slot WG comprises a slot waveguides whose rails are connected to two doped 70 nm thick silicon strips. The mode has a high overlap with the slot region.

quasi-TE polarized light. Four major classes of singlemode waveguides (WG) can be distinguished and are depicted in Fig. 2.3. The most common geometry is a fully etched waveguide, the so-called strip WG, Fig. 2.3(a), which has a height of 220 nm and a width of 450 nm. The light is strongly confined to the silicon core. Field discontinuities of the quasi-TE polarized mode increase the scattering loss due to sidewall roughness [6]. Due to recent technological improvements, propagation loss of these waveguides is typically in the order of (2…3) dB/cm or even below [3]. Strip WG allow for a high integration density and for nearly lossless bends (0.03 dB/90°) with a radius down to 5 µm [3]. Most of the components that are available for the SOI platform, like multimode interference couplers, are based on strip waveguides. Wherever propagation loss is a concern, shallow etched rib WG, Fig. 2.3(b), so-called "ridge" WGs can be used. These singlemode waveguides have a width of 700 nm and comprise an etch depth of 70 nm. The mode is less confined as compared to strip waveguides, but the propagation loss is significantly lower, i. e., losses as low as 0.3 dB/cm have been demonstrated [3]. Two other waveguide geometries have proven to be particularly interesting for the silicon-organic hybrid platform, where evanescent field interaction of the guided mode with an organic cladding is used to exploit the optical properties of the cladding material [26], see also Section 2.4. These waveguides are called slot WG, Fig. 2.3(c), and striploaded slot WG, Fig. 2.3(d). Both geometries strongly confine the light in a roughly 100 nm wide slot, which enables efficient interaction between light and cladding material. Slot WGs have previously been used for high-speed all-optical signal processing [43], [44] and for optically pumped lasing in silicon [45], while striploaded slot waveguides play a key role for the realization of low-power EO modulators [26], [31], [40], [46], [47] as described in Section 4.2.

2.3 Coupling Optical Fibers to SOI Waveguides

Despite the numerous advantages of the SOI platform, the realization of strongly guiding silicon waveguides leads to challenges when coupling the photonic integrated circuit (PIC) to glass fibers or other weakly guiding waveguide structures. This is due to the large mode mismatch between the waveguides. An important parameter of a waveguide is the relative refractive index difference between waveguide core and cladding material,

$$\Delta = \frac{n_{\text{core}}^2 - n_{\text{clad}}^2}{2n_{\text{core}}^2} \tag{2.25}$$

Mode fields of exemplary material platforms are compared to the mode of a typical SOI strip waveguide in Fig. 2.4. A weakly guided mode of a polymer waveguide with a relative refractive index difference of $\Delta = 0.001$ is depicted leftmost. The mode is matched to the mode of a standard singlemode fiber (SMF) and has a mode field diameter (MFD) of 10 µm. This MFD is significantly larger than the one of a silicon strip waveguide, which is only about 0.5 µm in size, resulting in very low coupling efficiencies between silicon waveguides and fibers. Different approaches have been investigated in order to improve the coupling efficiency between a fiber and a silicon PIC. The most

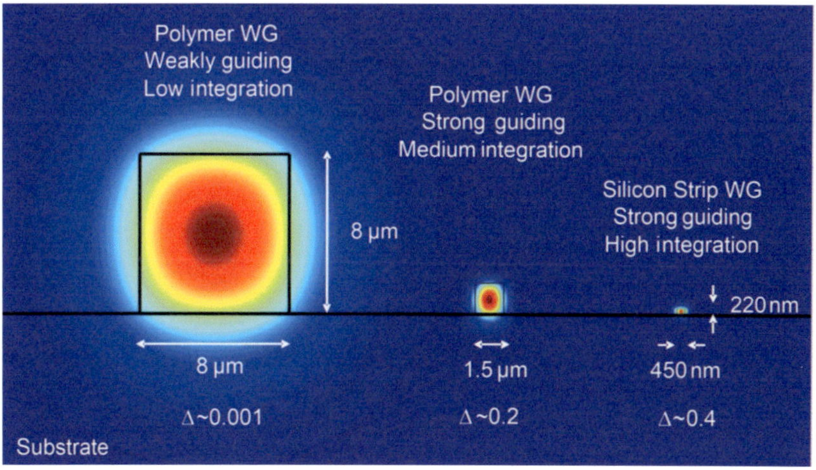

Fig. 2.4 Mode fields of three WG on exemplary material platforms are depicted: Weakly guiding polymer waveguide, strongly guiding polymer waveguide, and strongly guiding silicon waveguide. A large mode field mismatch between strongly and weakly guiding structures can be observed.

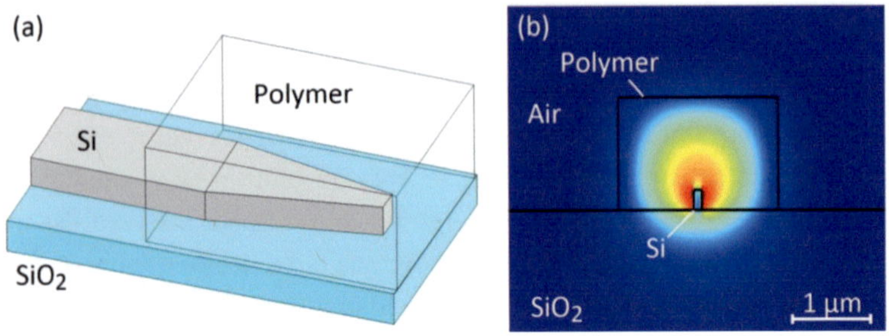

Fig. 2.5 Inverted silicon taper surrounded by a polymer waveguide as spot-size converter for butt-coupling. (a) Schematic of the inverted taper structure. The silicon waveguide width is tapered down from 450 nm to 100 nm. (b) Cross-section at the end of the taper and simulated mode profile. The mode is no longer guided by the narrow silicon waveguide. It is now guided by the polymer waveguide.

straightforward approach is the use of so-called lensed fibers that focus the fiber beam down to a MFD of (2…3) µm. Still, this MFD is much larger than those of strip WGs, and coupling efficiency between a cleaved strip WG and a lensed fiber is only around –5 dB. In addition, the approach is particularly prone to misalignment between fiber and silicon WG and requires precise nano-positioning stages for stable coupling.

2.3.1 Spot-Size Converter based on Inverted Silicon Taper

Coupling efficiency can be increased while simultaneously increasing alignment tolerances by tapering down the width of the strip WG to 100 nm and cladding the taper with a polymer WG or a SiO_2 WG as depicted in Fig. 2.5. This way, the mode field expands into the cladding and is finally guided by the surrounding polymer or glass waveguide only. The MFD is increased to roughly 2 µm, resulting in a coupling loss between silicon strip waveguide and lensed fiber as low as 1 dB [11]. Alignment tolerances are slightly relaxed. A coupling penalty of roughly 1 dB results from an axis offset of 1.5 µm. Drawbacks of this approach are the need of expensive lensed fibers and of cleaved or polished WG facets.

2.3.2 Surface Grating Couplers for Vertical Fiber-Chip Coupling

Alternatively, an out-of-plane fiber can couple to the silicon PIC, as depicted in Fig. 2.6(a). For a qualitative understanding the incident beam is considered to be a plane wave that propagates though a medium with refractive index n_1.

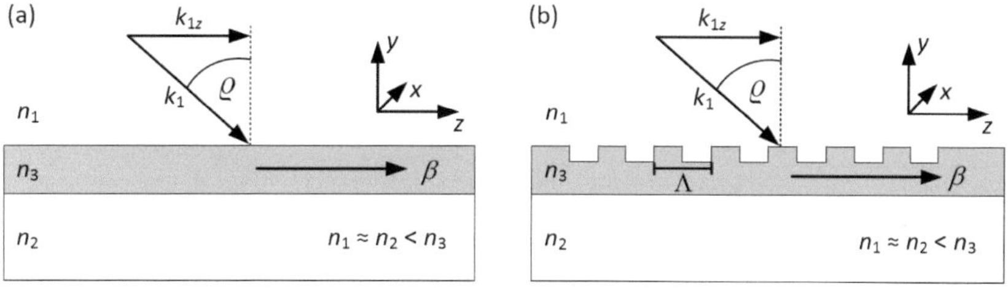

Fig. 2.6 Out-of-plane coupling between fiber and PIC. (a) Out-of-plane illumination without special coupling geometry. The propagating mode cannot be excited due to $\beta > k_1$. (b) Illumination of a grating coupler. The grating provides momentum to achieve phase matching of incoming wave and guided wave.

The wave vector of the incident beam is $\vec{k}_1 = \vec{k}_0 n_1$, and $\vec{k}_0$ is the wave vector in vacuum. In the silicon waveguide the light propagates in z-direction with propagation constant β. For an out-of-plane excitation, phase matching between incident wave and guided wave is required,

$$k_{1,z} = k_{0,z} n_1 = \beta = k_0 n_{\text{eff}} \,. \tag{2.26}$$

Here, $k_{1,z}$ is the z-component of the wave vector of the incident beam, and $n_{\text{eff}} = \beta / k_0$ is the effective refractive index of the guided mode in the WG. Since the refractive index n_3 of the waveguide core is much larger than the indices of substrate and cladding, the relation $n_1 < n_{\text{eff}}$ holds, which means that Eq. (2.26) cannot be fulfilled.

For enabling an out-of-plane excitation, a grating structure can be used, as depicted in Fig. 2.6(b). The grating of period Λ has a momentum K, which modifies the phase matching condition to

$$k_{1,z} = \beta + mK, \qquad K = \frac{2\pi}{\Lambda}. \tag{2.27}$$

The integer m is called diffraction order. Since $0 < k_{1,z} < \beta$ holds, this condition can only be fulfilled for negative diffraction orders. Due to time inversion symmetry of light propagation, this equation also holds when coupling light from the PIC back to a fiber. In the following, we will restrict the discussion to the case when light is coupled from the PIC to the fiber.

A graphical illustration of Eq. (2.27) is shown in Fig. 2.7. The wave vector diagram depicts the wave vector β (blue arrow) of the guided wave and the various diffraction orders (red arrows). Each groove of the grating radiates a cylindrical wave. Constructive interference of these waves only occurs in discrete directions that fulfil Eq. (2.27). The radii of the circles in Fig. 2.7 repre-

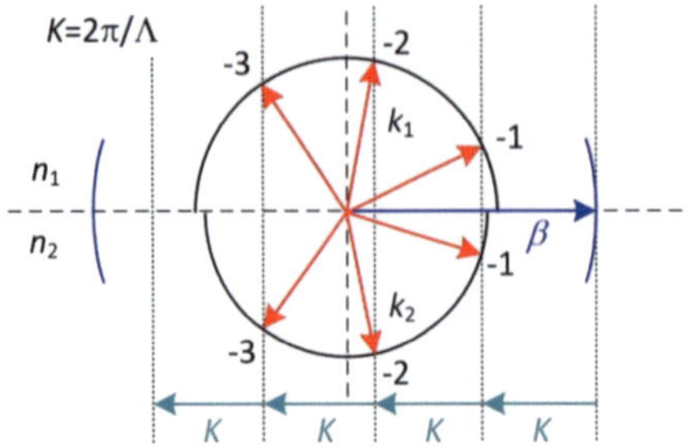

Fig. 2.7 Wave vector diagram. The diagram is a graphical representation of Eq. (2.27). Crossings of circles and dashed vertical lines mark radiation directions, for which phase matching of out-coupled wave and guided wave is fulfilled. The radius of a circle is equal to the length of the wave vector in the respective medium. The depicted diagram corresponds to a grating that supports 3rd-order diffraction in direction of the fiber.

sent the length of the wave vector of a plane wave in the respective medium. The dotted lines mark the values of $k_{1,z}$ (or $k_{2,z}$) in Eq. (2.27), which correspond to the z-components of a propagating plane wave in medium n_1 (or n_2). Phase matching is achieved wherever a circle intersects with a dotted vertical line. The example in Fig. 2.7 shows a grating that radiates three diffraction orders into the substrate, and three diffraction orders into the cladding. This is rather disadvantageous for the realization of an efficient fiber-chip coupler, because the radiated power is distributed among 6 diffraction orders.

In order to achieve efficient fiber-chip coupling, the grating period Λ needs to be chosen such that only one diffraction order is radiated into the cladding (besides the radiation into the substrate). From Fig. 2.7 it can be seen that this can be achieved by choosing a period $\Lambda = 2\pi / K$ that fulfils

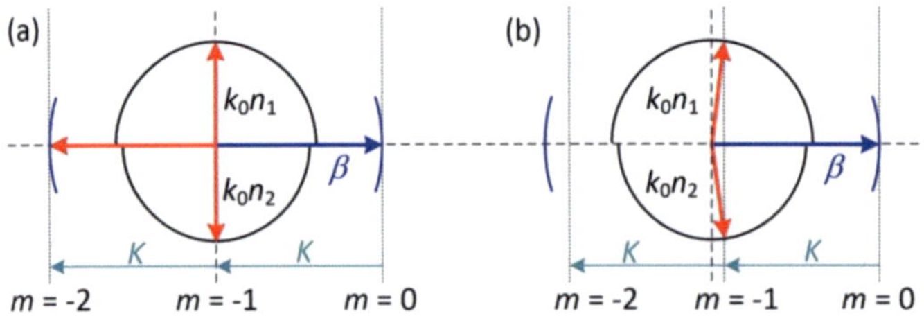

Fig. 2.8 Phase matching using a lattice structure with period Λ. (a) The lattice period has been chosen such that the first-order diffraction is vertical to the chip surface. This leads to the existence of a second-order diffraction, which is a back-reflection into the waveguide. (b) A slight detuning of the lattice period suppresses the second-order diffraction, but results in a slightly angled emission of the first-order diffraction with respect to the surface normal.

$(\beta + k_1)/2 < K < \beta$. Wave vector diagrams of two exemplary grating structures are depicted in Fig. 2.8. A second-order Bragg grating, $K = \beta$, is depicted in Fig. 2.8(a). For this grating type the first-order diffraction is radiated vertically to the PIC surface. The also existing second-order diffraction is a detrimental back-reflection into the waveguide. Therefore second-order Bragg gratings are usually avoided. The detrimental second-order diffraction can be eliminated by a small increase of the grating period, as depicted in Fig. 2.8(b). However, this comes at the price of a slight detuning of the out-coupling angle with respect to the chip surface normal.

A schematic of such a first-order diffraction grating coupler is depicted in Fig. 2.9(a). The grating has a width of 10 µm for matching the mode size of a standard singlemode fiber. A linear adiabatic transition is used to couple the strip WG to the grating coupler. The grating has a period of 630 nm, a groove width of 315 nm, and a groove etch depth of 70 nm as was previously reported

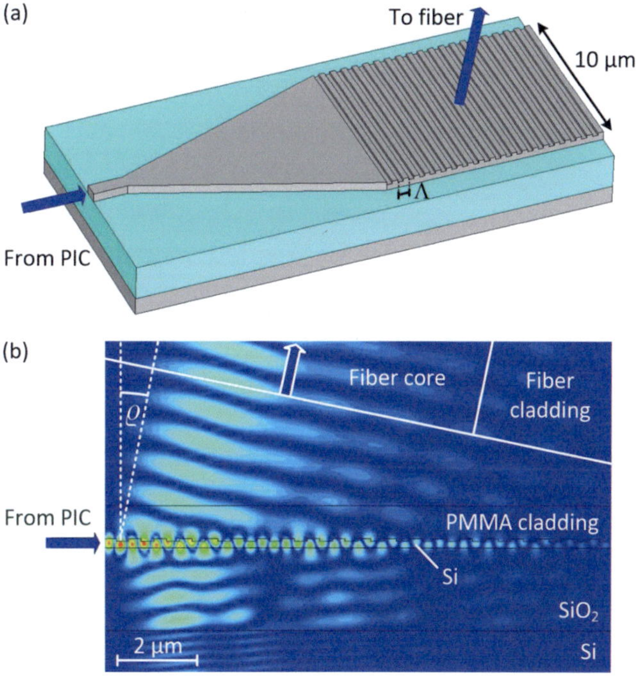

Fig. 2.9 Surface grating coupler for out-of-plane fiber-chip coupling. (a) Schematic of a grating coupler. (b) Simulated beam profile of a silicon grating coupler. Light is coupled out of plane at an angle ϱ of roughly 10° with respect to the surface normal.

in [48]. A simulation of the field propagation in the grating is depicted in Fig. 2.9(b). The input light from the PIC at a wavelength of 1550 nm is coupled out of plane into the optical fiber at an angle $\varrho = 10°$ with respect to the surface normal. The grating comprises a first order diffraction ($m = -1$) into the cladding and into the substrate. Therefore about half of the diffracted light is lost to the substrate.

The wave vector diagram Fig. 2.8(b) indicates only which diffraction orders exist, but not the amount of power that is radiated, and not the coupling efficiency to a singlemode fiber. The coupling efficiency η of the grating coupler denotes the fraction of power that is coupled from the grating to the fiber or vice versa. It is given as a product of the directionality D and the power coupling coefficient $\kappa^{(P)}$, see Eq. (B.20), between the emitted beam profile and the fiber mode,

$$\eta = D\kappa^{(P)} = \frac{P_{\text{output, fiber}}}{P_{\text{input, GC}}}. \tag{2.28}$$

The directionality factor is defined as the fraction of guided power power in the silicon WG, $P_{\text{input, GC}}$, that is radiated upwards in direction of the fiber,

$$D = \frac{P_{\text{up}}}{P_{\text{input, GC}}}. \tag{2.29}$$

The power coupling coefficient $\kappa^{(P)}$, Eq. (B.20), takes the form

$$\kappa^{(P)} = \frac{\left| \int_{-\infty}^{+\infty} \left(\vec{E}_{\text{GC}} \times \vec{\mathcal{H}}_{\text{fib}}^{*} + \vec{\mathcal{E}}_{\text{fib}}^{*} \times \vec{H}_{\text{GC}} \right) \vec{e}_z \, \mathrm{d}A \right|^2}{\int_{-\infty}^{+\infty} \left(\vec{E}_{\text{GC}} \times \vec{H}_{\text{GC}}^{*} + \vec{E}_{\text{GC}}^{*} \times \vec{H}_{\text{GC}} \right) \vec{e}_z \, \mathrm{d}A \cdot \int_{-\infty}^{+\infty} \left(\vec{\mathcal{E}}_{\text{fib}} \times \vec{\mathcal{H}}_{\text{fib}}^{*} + \vec{\mathcal{E}}_{\text{fib}}^{*} \times \vec{\mathcal{H}}_{\text{fib}} \right) \vec{e}_z \, \mathrm{d}A}, \tag{2.30}$$

where $\vec{E}_{\text{GC}}$, $\vec{H}_{\text{GC}}$ denote the electric and magnetic field vectors of the upwards radiated beam, and $\vec{\mathcal{E}}_{\text{fib}}$, $\vec{\mathcal{H}}_{\text{fib}}$ stand for the respective transverse mode fields of the fiber. The integral in the numerator is called overlap integral.

The coupling efficiency of such a grating coupler is mostly limited by its directionality, resulting in coupling losses of 3 dB to 5 dB [48]. The directionality of a grating coupler can be strongly enhanced by either breaking the structure symmetry, e. g., by adding a poly-silicon overlay [49], or by using a metallic mirror underneath the buried oxide that reflects the downward diffracted light [50], [51], [52]. The highest reported coupling efficiency amounts to -0.6 dB and was realized with a "chirped" grating optimized for a Gaussian

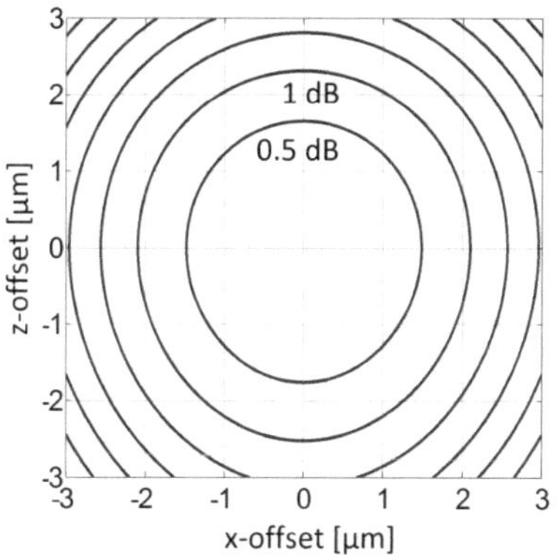

Fig. 2.10 Alignment tolerances of the grating coupler from Fig. 2.9 are depicted. The lines represent power penalties in steps of 0.5 dB beginning in the center of the image. A misalignment of the fiber by $\pm 2\,\mu m$ results in less than 1 dB additional coupling loss.

beam profile in combination with a metallic back-reflector [13]. Surface grating couplers support wafer-scale testing of devices, which makes them particularly interesting for mass production. They feature relaxed alignment tolerances, see Fig. 2.10, where a lateral misalignment of ±2 µm results in a power penalty of less than 1 dB. Despite of these advantages, the vertical arrangement of fiber and chip makes packaging of the PIC challenging. In general, an in-plane arrangement of fiber and chip is preferred by the industry. To overcome this issue, a specially polished fiber has been developed that allows an in-plane fiber arrangement using surface grating couplers, as reported in Section 4.

2.4 Electro-Optic Mach-Zehnder Modulator Realized on the Silicon-Organic Hybrid (SOH) Platform

Building PIC on the SOI platform offers many advantages such as high integration density and compatibility with mature CMOS processes. However, the SOI platform poses also issues when targeting at monolithic transceiver units for telecom and datacom: The indirect band gap and the inversion symmetry of crystalline silicon prevent efficient lasing and modulation via the linear EO effect (Pockels effect), respectively. The silicon-organic hybrid (SOH) platform overcomes these limitations by combining the benefits of the SOI platform with the advantage of tailoring the optical properties of organic materials. Consequently, optically pumped SOH lasers were demonstrated with a silicon slot waveguide that is filled with a dye-doped organic material [45].

This section focusses on a concept for realizing energy-efficient, high-speed SOH electro-optic modulators. These modulators are based a strip-loaded slot waveguide, Fig. 2.3(d), clad by an EO organic material. The basic operation principle, design aspects, bandwidth limitations, and the fabrication of the device are discussed.

2.4.1 Modulator Concept

A schematic of an SOH Mach-Zehnder modulator (MZM) is depicted in Fig. 2.11. The MZM consists of two SOH phase modulators that are driven in push-pull by a single coplanar transmission line in GSG (ground, signal, ground) configuration. Grating couplers are used for fiber-chip coupling, and 2×2 multimode interference couplers (MMI) are used as power splitters and combiners. Strip waveguides, see Fig. 2.3(a), are used as access waveguides to connect grating couplers, MMIs, and phase modulators. Strip-to-slot waveguide mode converters are used as interconnects between strip and slot waveguides. A cross-sectional view of the modulator section is displayed in Fig. 2.12(a). A schematic of an SOH phase modulator along with the electric field of the fundamental quasi-TE mode is depicted in Fig. 2.12(b). The phase modulator consists of a striploaded slot waveguide that comprises two silicon rails separated by a 100 nm wide slot. The waveguide is covered and the slot is filled with an EO organic material that is deposited by spin-coating. The rails of the slot waveguide are electrically connected to the RF transmission line by

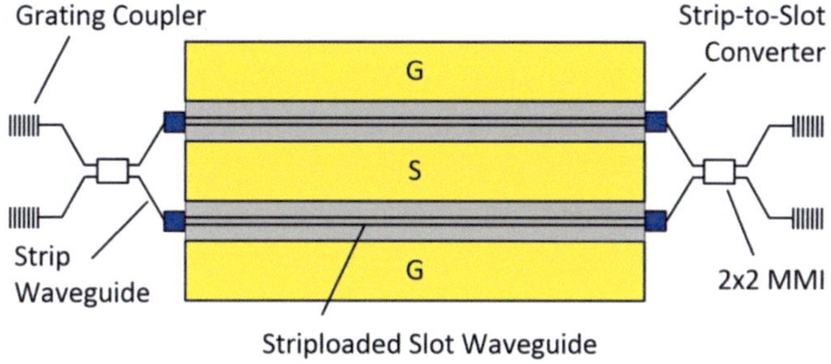

Fig. 2.11 Schematic of an SOH Mach-Zehnder modulator. The MZM section consists of two striploaded slot waveguides that are driven by one coplanar transmission line in a ground-signal-ground (GSG) configuration. Strip waveguides are used as access waveguides. Strip-to-slot mode converters couple strip and slot waveguides. Power splitters and combiners are realized as 2×2 multimode interference couplers (MMI).

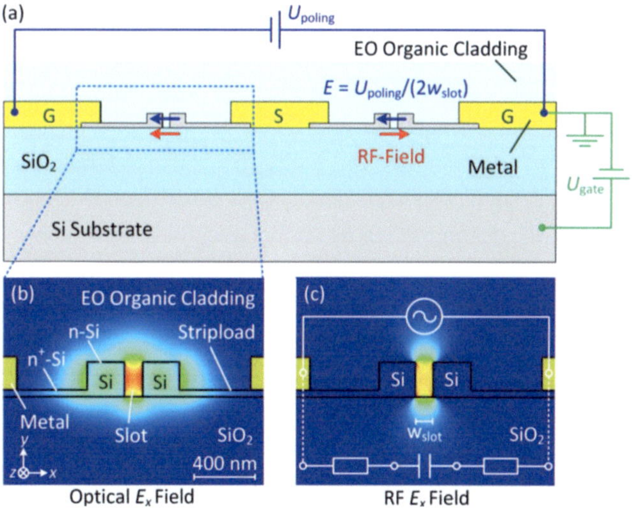

Fig. 2.12 Silicon-organic hybrid Mach-Zehnder modulator. (a) Cross-section of the MZM. The two phase modulators are driven in push-pull by a single coplanar transmission line in GSG (ground, signal, ground) configuration. The EO organic cladding is poled by applying a voltage U_{poling} across the floating ground electrodes of the transmission line. A gate voltage U_{gate} between substrate and transmission line increases the bandwidth of the modulator [31] by formation of an electron accumulation layer in the silicon striploads. (b) Simulated optical mode in the phase modulator section. The phase modulator consists of a striploaded slot waveguide that is covered and filled with an EO organic material. (c) Simulated RF-field in the phase modulator section. The modulation field is confined to the slot region and has a large overlap with the optical mode. Modified after [J13]. Copyright © 2014 IEEE.

70 nm thick *n*-doped silicon strips (so-called stripload). The strips act, together with the slot waveguide capacitance, as an RF load. A voltage applied to the transmission line drops across the narrow slot, resulting in a strong modulation field, see Fig. 2.12(c), that has a large overlap with the guided optical mode, Fig. 2.12(b). For increasing the stripload conductivity a gate field can be applied between transmission line and silicon substrate as depicted in Fig. 2.12(a). This way a quasi-two-dimensional electron accumulation layer is formed in the stripload at the interface between stripload and BOX. Details on field-induced electron accumulation can be found in Ref. [31].

For high-modulation efficiency, an organic material with a large macroscopic EO coefficient r_{33} is essential. The macroscopic EO coefficient depends on the orientation of the chromophore ensemble with respect to the modulation field, see Eq. (2.24). After deposition of the organic cladding, the electo-optic coefficient r_{33} vanishes due to the random orientation of the dipolar chromophores. For achieving a non-centrosymmetric orientation (acentric order), poling of the material is required. Poling is defined as molecular re-orientation induced by an electric field while heating the material to the glass-transition temperature T_g. After molecular alignment, the chip is cooled down to room temperature while maintaining the poling field, thereby freezing the acentric order of the chromophores and the macroscopic EO coefficient. In a single-drive MZM this is accomplished by applying a poling voltage U_{poling} across the two ground electrodes of the RF transmission line as depicted in Fig. 2.12(a). This way, half of the poling voltage drops across each silicon slot resulting in an antisymmetric re-orientation of the dipolar chromophores (indicated by blue arrows) with respect to the modulation field of the transmission line (indicated by red arrows). As a consequence, the two phase modulators provide opposite phase shifts (push-pull operation, see Appendix E.1). According to Eq. (E.7) the output field amplitude is related to the phase difference $\Delta\varphi = \varphi_2 - \varphi_1$ in the two arms of the MZM leading to the well-known cos-characteristic of the output field

$$E_{\text{out}} \propto \cos(\Delta\varphi). \tag{2.31}$$

The phase shift in each arm of the Mach-Zehnder is a result of Pockels effect interaction between optical electric field and the applied modulation field E_{m}, see Section 2.1.3.

According to Eq. (C.17) the phase shift $\delta\varphi$ in one arm of the MZM is given by

$$\begin{aligned}\delta\varphi &= -\delta\beta L \\ &= \frac{\varepsilon_0 \omega_c L}{4\mathcal{P}_0(\omega_c)} \iint_{-\infty}^{\infty} n_0^4(\vec{r}) r_{33}(\vec{r}) E_m(\vec{r},t,\omega_m \approx 0) \vec{\mathcal{E}}_0(x,y,\omega_c) \vec{\mathcal{E}}_0(x,y,\omega_c) \,\mathrm{d}x\,\mathrm{d}y,\end{aligned} \tag{2.32}$$

where L is the length of the modulator, ω_c is the optical carrier frequency, $\vec{\mathcal{E}}_0$ is the electric field of the optical fundamental mode of the slot waveguide, $\mathcal{P}_0$ is the power of the optical fundamental mode, and r_{33} is the EO coefficient of the organic cladding. In the case of a striploaded slot waveguide we may further simplify this expression by assuming that the organic material is only poled within the volume of the slot region V_{slot}:

$$r_{33}(\vec{r}) = \begin{cases} r_{33} & \forall\, \vec{r} \in V_{slot} \\ 0 & \forall\, \vec{r} \notin V_{slot}. \end{cases} \tag{2.33}$$

This reduces the integration domain of the integral to the slot region. Assuming a uniform external modulation field E_m in the slot simplifies Eq. (2.32) to

$$\delta\varphi = \frac{\varepsilon_0 \omega_c L}{4\mathcal{P}_0(\omega_c)} n_{slot}^4 r_{33} E_m \iint_{A_{slot}} \vec{\mathcal{E}}_0(x,y,\omega_c) \vec{\mathcal{E}}_0^*(x,y,\omega_c) \,\mathrm{d}x\,\mathrm{d}y. \tag{2.34}$$

By defining the field interaction factor Γ

$$\Gamma = \frac{n_{slot}}{Z_0} \frac{\int_{A_{slot}} \left|\vec{\mathcal{E}}_0\right|^2 \mathrm{d}A}{2\mathcal{P}_0(\omega_c)} \tag{2.35}$$

Eq. (2.34) can be rewritten in a more compact form

$$\delta\varphi = \frac{\pi}{\lambda_c} \Gamma n_{slot}^3 r_{33} L E_m. \tag{2.36}$$

Here, the following expressions for the free-space impedance Z_0 and for the speed of light c have been used:

$$Z_0 = \sqrt{\frac{\mu_0}{\varepsilon_0}}, \qquad c = \frac{1}{\sqrt{\varepsilon_0 \mu_0}}. \tag{2.37}$$

The π-voltage U_π of the push-pull MZM is defined as the voltage that induces a relative phase shift of π between the two arms of the Mach-Zehnder interferometer. Therefore the required phase shift for switching between maximum and minimum of transmission is $+\pi/2$ in the two phase shifters of the push-pull modulator, respectively.

This leads us to the following expression:

$$\delta\varphi = \frac{\pi}{2} = \frac{\pi}{\lambda_c} \Gamma n_{\text{slot}}^3 r_{33} L E_{\text{m}} = \frac{\pi}{\lambda_c} \Gamma n_{\text{slot}}^3 r_{33} L \frac{U_\pi}{w_{\text{slot}}}. \tag{2.38}$$

Thus, the U_π-voltage of the push-pull MZM is given by

$$U_\pi = \frac{w_{\text{slot}} \lambda_c}{2 L r_{33} \Gamma n_{\text{slot}}^3}. \tag{2.39}$$

The drive voltage of the SOH MZM thus depends in part on the properties of the organic material in the slot (r_{33}, n_{slot}, Γ), and in part on the geometrical properties of the device (w_{slot}, Γ, L).

In the following, we will discuss the optimization of the geometrical properties of the MZM for realizing devices with low drive voltage.

2.4.2 Modulator Dimensions

According to Eq. (2.39) a low drive voltage of the SOH MZM can be achieved by choosing a large device length L, by designing the geometry of the strip-loaded slot waveguide for highest field interaction factor Γ, and by choosing a small slot width w_{slot}. However, as will be discussed in Section 2.4.3, a large device length and a narrow slot width both lead to limitations of the modulation bandwidth. A long device is more prone to walk-off between optical signal and RF signal. A narrow slot width increases the device capacitance which results in a small RC cut-off frequency. Therefore compromises have to be made when aiming at low drive voltage and high bandwidth simultaneously.

The field interaction factor Γ, see Eq. (2.35), of a striploaded slot waveguide with a stripload height of 70 nm is depicted in Fig. 2.13 as a function of rail width and slot width for an organic cladding material with refractive index of 1.8. The highest field interaction factor of $\Gamma = 0.22$ is achieved for a slot width of 80 nm and a rail width of 240 nm. Smaller slot widths are impractical for fabrication, even when using eBeam lithography.

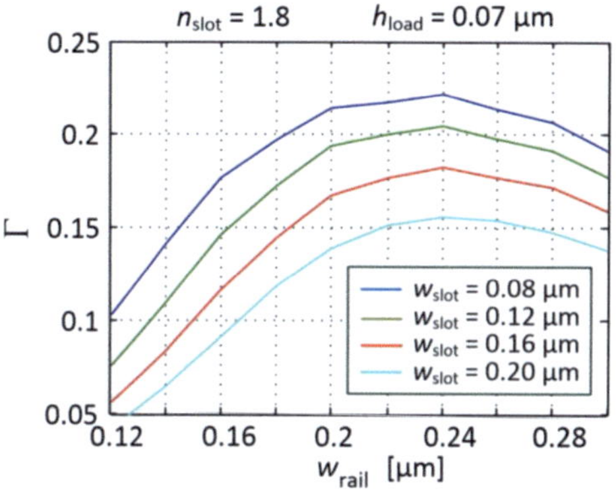

Fig. 2.13 Field interaction factor Γ calculated for different rail and slot widths and for a cladding refractive index of 1.8.

2.4.3 Bandwidth

The bandwidth of the SOH MZM is limited by three factors: Velocity mismatch between light and microwave signal, microwave propagation loss, and RC cut-off. In this subsection an analytic expression is derived for estimating the bandwidth of the modulator.

An expression that describes the frequency-dependent sinusoidal phase modulation $\delta\varphi$ of a phase modulator with regard to microwave loss and velocity mismatch is given in Ref. [39],

$$\delta\varphi(f_{\mathrm{m}},t) = \delta\varphi_0 \frac{1}{L}\int_0^L \mathrm{e}^{-\alpha_{\mathrm{m}}(f_{\mathrm{m}})z/2} \cos\left(2\pi f_{\mathrm{m}}\left(\frac{\Delta t_{\mathrm{o,m}}}{L}z - t\right)\right)\mathrm{d}z\,, \tag{2.40}$$

where $\delta\varphi_0$ is a scaling factor which is proportional to the modulating field, L is the length of the modulator, and α_{m} is the power attenuation coefficient of the transmission line. The time difference $\Delta t_{\mathrm{o,m}}$ expresses a phase delay between the sinusoidal optical and microwave fields. However, for a practical application where the modulation deviates from a sinusoidal, it is the group delay difference of optical and microwave signal which counts. This "walk-off" is then given by the expression

$$\Delta t_{\mathrm{o,m}} = \left|\frac{L}{v_{\mathrm{g}}^{\mathrm{m}}} - \frac{L}{v_{\mathrm{g}}^{\mathrm{o}}}\right| = \frac{L}{c}\left|n_{\mathrm{g}}^{\mathrm{m}} - n_{\mathrm{g}}^{\mathrm{o}}\right|, \tag{2.41}$$

where v_g^m and v_g^o are the group velocities of the microwave signal and of the optical signal, respectively, and n_g^m, n_g^o denote the respective group indices.

Assuming matched group velocities $\Delta t_{o,m} = 0$, the amplitude of the phase $\delta\varphi(f_m)$ reads

$$\delta\varphi(f_m) = \delta\varphi_0 \frac{1}{L}\int_0^L e^{-\alpha_m(f_m)z/2}\mathrm{d}z = \delta\varphi_0 \frac{L_{\mathrm{eff}}(f_m)}{L}. \tag{2.42}$$

Here, we introduced a frequency-dependent effective device length L_{eff}

$$L_{\mathrm{eff}}(f_m) = \frac{2}{\alpha_m(f_m)}\left[1 - e^{-\alpha_m(f_m)L/2}\right]. \tag{2.43}$$

For skin effect dominated microwave attenuation the frequency-dependency of the power attenuation factor α_m can be written as (Eq. (D.16))

$$\alpha_m(f_m) = a\sqrt{f_m}, \quad a = \mathrm{const.} \tag{2.44}$$

In the limit of small data rates, the striploaded slot waveguide of the modulator acts as a lumped element and can be interpreted as an RC-circuit. The lumped-element equivalent circuit of a lumped phase modulator, connected to a source with open circuit voltage U_0 and internal series resistor $R_s = 50\,\Omega$ is depicted in Fig. 2.14(a). The phase modulator itself is represented by the silicon stripload resistor R_{load} and by the slot capacitor C_{slot}, connected in series. A resistor $R_t = 50\,\Omega$ is used to terminate the signal. The phase shift in the phase modulator is proportional to the voltage U_c that drops across the slot

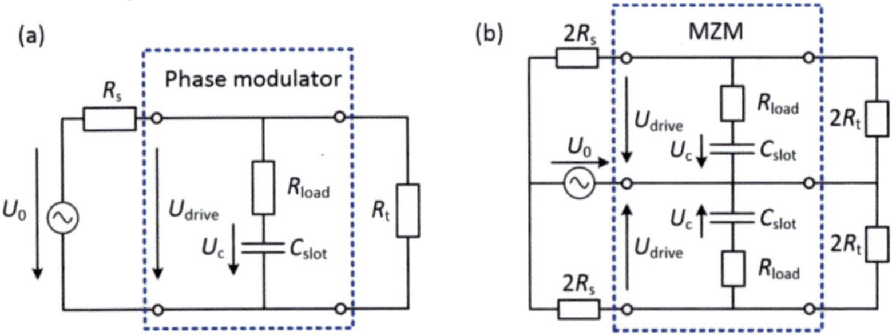

Fig. 2.14 Lumped-element equivalent circuits of an SOH phase modulator and of a single-drive SOH push-pull MZM, both terminated and connected to a RF voltage source. (a) The phase modulator is represented only by a series resistor R_{load} and by a slot capacitor C_{slot}. (b) The MZM is represented by two parallel slot capacitors C_{slot}, each connected to a series resistor R_{load}. Further symbols: Open circuit source voltage U_0, drive voltage U_{drive}, capacitor voltage U_c, internal source resistor $R_s = 50\,\Omega$, terminating resistor $R_t = 50\,\Omega$.

capacitor, $\delta\varphi \propto U_\mathrm{c}$. The capacitor voltage U_c is frequency-dependent and can be expressed by

$$\begin{aligned}|U_\mathrm{c}(f_\mathrm{m})| &= |U_\mathrm{drive}(f_\mathrm{m})|\frac{|Z_\mathrm{slot}(f_\mathrm{m})|}{|R_\mathrm{load}+Z_\mathrm{slot}(f_\mathrm{m})|}\\ &= |U_0|\frac{R_\mathrm{t}|R_\mathrm{load}+Z_\mathrm{slot}|}{|R_\mathrm{s}(R_\mathrm{t}+R_\mathrm{load}+Z_\mathrm{slot})+R_\mathrm{t}(R_\mathrm{load}+Z_\mathrm{slot})|}\frac{|Z_\mathrm{slot}|}{|R_\mathrm{load}+Z_\mathrm{slot}|}\\ &\overset{R_\mathrm{s}=R_\mathrm{t}}{=} |U_0|\frac{|R_\mathrm{load}+Z_\mathrm{slot}|}{|R_\mathrm{s}+2R_\mathrm{load}+2Z_\mathrm{slot}|}\frac{|Z_\mathrm{slot}|}{|R_\mathrm{load}+Z_\mathrm{slot}|}, \text{ with } Z_\mathrm{slot}=\frac{1}{\mathrm{j}2\pi f_\mathrm{m}C_\mathrm{slot}}.\end{aligned} \tag{2.45}$$

For RF wavelengths that are much larger than the device length the walk-off between optical wave and microwave can be neglected. In this case we combine the lumped RC-circuit model in Eq. (2.45) with the model for microwave attenuation, Eq. (2.42),

$$\delta\varphi(f_\mathrm{m}) = \delta\varphi_0\frac{2|R_\mathrm{load}+Z_\mathrm{slot}(f_\mathrm{m})|}{|R_\mathrm{s}+2R_\mathrm{load}+2Z_\mathrm{slot}(f_\mathrm{m})|}\frac{|Z_\mathrm{slot}(f_\mathrm{m})|}{|R_\mathrm{load}+Z_\mathrm{slot}(f_\mathrm{m})|}\frac{L_\mathrm{eff}(f_\mathrm{m})}{L}. \tag{2.46}$$

The factor 2 is added such that the phase shift is $\delta\varphi(0)=\delta\varphi_0$ at a modulation frequency $f_\mathrm{m}=0$, see Eq. (2.44). The modulation bandwidth of a phase modulator can be specified by inserting the phase modulator into a Mach-Zehnder interferometer and by operating the device as an approximately linear intensity modulator (i. e., operation at the quadrature point, see Appendix E). The analytic expression for small signal intensity modulation $\delta I \propto \delta\varphi$ is given by

$$\frac{\delta I(f_\mathrm{m})}{\delta I_0} \approx \frac{2|R_\mathrm{load}+Z_\mathrm{slot}(f_\mathrm{m})|}{|R_\mathrm{s}+2R_\mathrm{load}+2Z_\mathrm{slot}(f_\mathrm{m})|}\frac{|Z_\mathrm{slot}(f_\mathrm{m})|}{|R_\mathrm{load}+Z_\mathrm{slot}(f_\mathrm{m})|}\frac{L_\mathrm{eff}(f_\mathrm{m})}{L}. \tag{2.47}$$

This relation thus holds for a *terminated MZM in which a single phase modulator is driven.*

The situation is different when driving a *terminated single-drive MZM, in which two phase modulators are driven in push-pull*, see Fig. 2.12. This case is depicted by the lumped-element equivalent circuit in Fig. 2.14(b). Here, the capacitor voltage can be expressed by

$$|U_\mathrm{c}(f_\mathrm{m})| \overset{R_\mathrm{s}=R_\mathrm{t}}{=} |U_0|\frac{|R_\mathrm{load}+Z_\mathrm{slot}(f_\mathrm{m})|}{|2R_\mathrm{s}+2R_\mathrm{load}+2Z_\mathrm{slot}(f_\mathrm{m})|}\frac{|Z_\mathrm{slot}(f_\mathrm{m})|}{|R_\mathrm{load}+Z_\mathrm{slot}(f_\mathrm{m})|}, \tag{2.48}$$

and the resulting linear small-signal intensity modulation (at quadrature) is

$$\frac{\delta I(f_\mathrm{m})}{\delta I_0} \approx \frac{2|R_\mathrm{load}+Z_\mathrm{slot}(f_\mathrm{m})|}{|2R_\mathrm{s}+2R_\mathrm{load}+2Z_\mathrm{slot}(f_\mathrm{m})|}\frac{|Z_\mathrm{slot}(f_\mathrm{m})|}{|R_\mathrm{load}+Z_\mathrm{slot}(f_\mathrm{m})|}\frac{L_\mathrm{eff}(f_\mathrm{m})}{L}. \tag{2.49}$$

Note that twice the source resistance R_s goes into Eq. (2.49) for the case of a push-pull MZM, in contrast to findings for the case of a MZM in which a single phase modulator is driven, see Eq. (2.47). Therefore, the frequency response of a single-drive push-pull MZM is expected to be worse than that of a single phase modulator at high frequencies.

The electro-optic (EO) response of a modulator is defined by its linear small-signal intensity modulation

$$S_{21,\mathrm{EO}}(f_\mathrm{m}) = 10\log_{10}\left(\frac{\delta I(f_\mathrm{m})}{\delta I(0)}\right), \tag{2.50}$$

and the EO 3 dB cut-off frequency $f_{3\mathrm{dB,EO}}$ is defined as the frequency where the magnitude of intensity modulation drops to half the value measured at low frequencies. The EO bandwidth is usually determined indirectly by measuring the electrical-optical-electrical (EOE) response.

The EOE response is measured by small-signal intensity modulation using a vector network analyzer (VNA), the device under test (DUT) and a high-speed photodiode (PD). The AC current $\delta i_\mathrm{PD}(f_\mathrm{m})$ of the PD and the measured voltage $\delta u_\mathrm{VNA}(f_\mathrm{m})$ at the $50\,\Omega$ input impedance of the VNA are proportional to the modulated optical input intensity δI,

$$\delta u_\mathrm{VNA} \propto \delta i_\mathrm{PD} \propto \delta I\,. \tag{2.51}$$

The measured EOE S_{21}-parameter of the device is defined as

$$\mathrm{S}_{21,\mathrm{EOE}}(f_\mathrm{m}) = 10\log_{10}\left(\frac{\delta u^2_\mathrm{VNA}(f_\mathrm{m})}{\delta u^2_\mathrm{VNA}(0)}\right) = 20\log_{10}\left(\frac{\delta I(f_\mathrm{m})}{\delta I(0)}\right). \tag{2.52}$$

Therefore, the EO 3 dB cut-off frequency $f_{3\mathrm{dB,EO}}$ corresponds to the EOE 6 dB cut-off frequency $f_{6\mathrm{dB,EOE}}$,

$$f_{3\mathrm{dB,EO}} = f_{6\mathrm{dB,EOE}}\,. \tag{2.53}$$

2.4.4 Fabrication Scheme

The process flow for the fabrication of the SOH modulators was developed in cooperation with AMO GmbH, Aachen, Germany. For fabrication, electron beam lithography is used. The process flow is depicted in Fig. 2.15. The substrate is an SOI wafer with a 220 nm thick silicon device layer and a 2 µm thick buried oxide. First, the doping regions are defined using positive tone photo resist (PR). Ion implantation is performed to achieve arsenic doping concentrations in the order of $(1...7)\times10^{17}\,\mathrm{cm}^{-3}$. After thermal annealing of the ion implant the slot waveguide is defined using hydrogen silsesquioxane (HSQ) negative tone resist, followed by a 150 nm silicon dry etch step. Next, a second HSQ layer is deposited on top of the existing HSQ structures to define the striploads. This way a self-aligned definition of the striploaded slot waveguide is established. A subsequent 70 nm silicon dry etch is performed to open the slots. Afterwards, the HSQ resist is stripped. Next, a second doping region is created to achieve a good Ohmic contact between silicon and metal. After doping and annealing, metal electrodes are deposited in a lift-off process. Finally, the EO organic cladding is deposited by spin-coating.

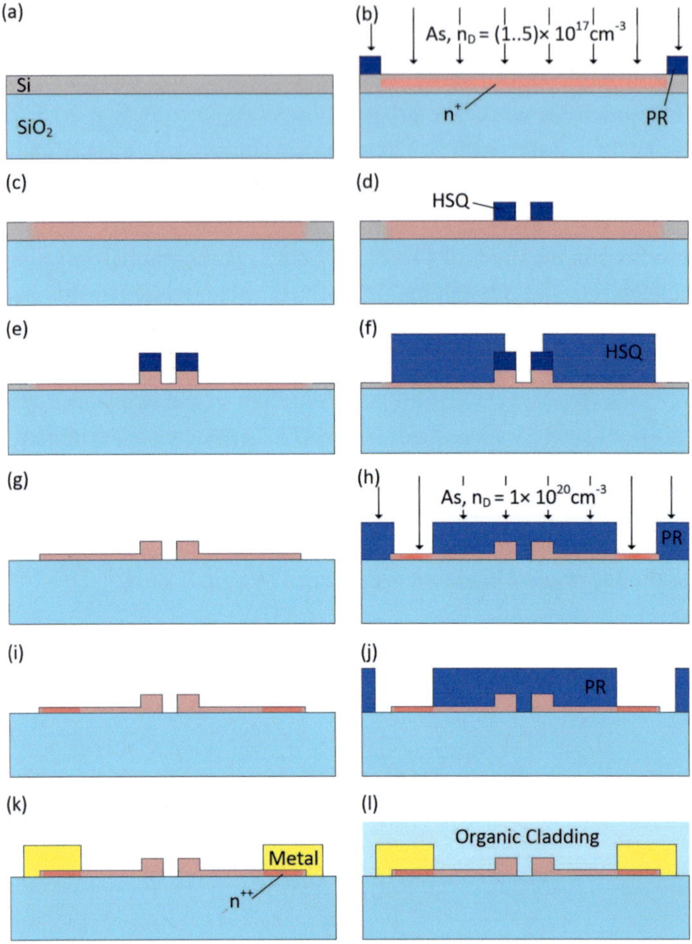

Fig. 2.15 Fabrication scheme of an SOH MZM. (a) SOI substrate with 220 nm thick device layer and 2 μm buried oxide. (b) Definition of doping regions and ion implant. (c) Photoresist (PR) removal and thermal annealing. (d) Waveguide definition using HSQ resist. (e) 150 nm silicon dry etch. (f) Definition of the striploads using HSQ resist. (g) 70 nm silicon dry etch and subsequent resist stripping. (h) Definition of region for high doping concentration and ion implant. (i) Resist stripping and annealing. (j) Electrode definition. (k) Lift-off metallization and subsequent resist stripping. (l) Spin-coating of the EO organic cladding.

3 Anisotropic Wet Etching of Waveguides

Low optical losses are a key requirement for any practical PIC. Fiber-chip coupling losses and optical propagation loss in the waveguide structures are two important contributors to the loss budget. In the following a novel fabrication scheme is introduced, based on anisotropic wet etching that potentially enables low propagation loss and low fiber-chip coupling losses.

The suitability of anisotropic wet etching for the fabrication of nanowire waveguides is demonstrated. Nanowire waveguides and grating couplers with trapezoidal cross-section are fabricated in a single etch step and characterized. Our simulations indicate that anisotropic wet etching allows for the fabrication of grating couplers with a Gaussian beam profile and 0.3 dB loss when coupling to standard singlemode fibers. In addition, we show that anisotropic wet etching in combination with a special lithographic mask layout can be exploited to fabricate nanowire structures with low-resolution optical lithography.

3.1 Introduction

Silicon-on-insulator (SOI) is expected a promising material platform for low-cost integrated photonic circuits. The high refractive index contrast between core and cladding material allows for high integration density, but makes the waveguide particularly prone to sidewall roughness [6]. The state of the art in fabrication of strongly guiding silicon nanowire waveguides is either based on electron beam lithography in combination with inductively coupled plasma etching [53] or relies on mature CMOS processes exploiting 193 nm deep-UV lithography [3]. Typical propagation losses of strip waveguides are in the order of 2 dB/cm, while recent technological improvements based on 193 nm deep-UV immersion lithography enabled losses as low as 0.7 dB/cm [4]. Despite all these technological improvements within the last decade, losses of silicon strip waveguides are still far larger than those of bulk silicon, where the material loss is smaller than 0.15 dB/cm [54]. For bridging cm-distances weakly guiding rib waveguides or multimode waveguides with losses of 0.3 dB/cm [3] are often preferred over strongly guiding strip waveguides.

Supplementing unconventional fabrication methods focus on waveguides with lowest sidewall roughness. One example is the so-called “etchless” waveguide, fabricated by inhomogeneous oxidation of a 500 nm thick SOI layer, where a low propagation loss of 0.3 dB/cm has been demonstrated [8]. The measured rms surface roughness is 0.3 nm. However, the mode of this waveguide is only weakly confined. The waveguide has a width of 1 µm and a height of only 80 nm, resulting in a low integration density. Furthermore, the fabrication process challenges the implementation of additional elements like gratings, modulators, and detectors.

Also post-processing methods like excimer laser induced reflow of the silicon waveguides has been studied [5] to reduce sidewall roughness. This method reduces an initial rms roughness of 14 nm to only 0.24 nm. From this roughness a propagation loss of 0.03 dB/cm is predicted by coupled mode theory [6]. However, this surface reformation technique is not scalable for mass production and the resulting round waveguide cross-section is not compatible to every component of an integrated circuit. Another post-processing method is atomic layer deposition (ALD) of TiO_2 onto the silicon waveguides [55]. ALD overgrowth results in a smoothening of the waveguide roughness. This way, losses of silicon slot waveguides have been reduced from 7 dB/mm to 7 dB/cm [55].

Besides plasma etching another common technique for structuring of silicon is anisotropic wet etching based on aqueous potassium hydroxide (KOH) or tetramethylammonium hydroxide (TMAH) solutions [56]. This technique is mainly used for the fabrication of microelectromechanical systems (MEMS), microfluidic channels or V-groove arrays. The fabrication of waveguide structures was so far limited to demonstrations of several µm wide rib waveguides [9], [57]. The applicability of this approach for the fabrication of nanowire waveguides is thus uncertain. The strong etching anisotropy of KOH in principle allows for nearly atomically flat waveguide sidewalls that could potentially result in low propagation losses even for narrow strip waveguides and slot waveguides [27]. Furthermore, the orientation of the {111}-crystal planes in silicon allows for the realization of novel waveguide shapes, e. g. with trapezoidal cross-section.

In the following, the suitability of wet etching for the fabrication of nanowire waveguides is demonstrated. Waveguide prototypes and grating couplers with trapezoidal cross-section are fabricated and charcterized. Simulations indicate that anisotropic wet etching allows for the fabrication of grating couplers with Gaussian beam profile and coupling efficiencies to standard singlemode fibers of –0.3 dB. In addition, it will be shown that crystal plane selectivity of wet etching in combination with a special lithographic mask layout can be exploited to fabricate smooth nano-wire structures though using low-resolution optical lithography.

3.2 Anisotropic Wet Etching and eBeam Lithography

Our fabrication method is based on anisotropic wet etching in a 40 % aqueous potassium hydroxide solution. The etch rate of silicon in a KOH solution depends strongly on the exposed crystal plane (see Glossary for the notation of crystal planes). The etch rate orthogonal to a {111}-crystal plane is about 300-times slower [56] as compared to other crystal planes. This favors the formation of stable {111}-planes during etching of crystalline silicon. However, this property also introduces design restrictions since all structures have to be aligned according to the {111}-crystal planes.

The arrangements of crystal planes for an SOI wafer with a (110)-orientation and for a wafer with (100)-orientation are depicted in Fig. 3.1(a) and (b), respectively. The waveguide structures are defined using a hard mask, either SiO_2 or Si_3N_4, which is aligned according to the {111}-crystal planes, as indicated by the dashed lines in Fig. 3.1(a,b). The resulting waveguide cross-sections are depicted in Fig. 3.1(c,d): The waveguide has a rectangular cross-section for a (110)-wafer, and it has a trapezoidal cross-section when using a (100)-wafer. The trapezoidal waveguide has a sidewall angle ϑ that corresponds to the angle between (100)-plane and (111)-plane,

$$\vartheta = \cos^{-1}\left(\frac{1}{\sqrt{3}}\right) = 54.74^\circ . \tag{3.1}$$

The bottom width of the waveguide w_{bottom} depends on the width of the top surface of the waveguide w_{top} and on its height h_{dev} by

$$w_{\text{bottom}} = w_{\text{top}} + 2h_{\text{dev}} / \tan\vartheta \tag{3.2}$$

In the following we focus on waveguide fabrication on SOI wafers with (100)-crystal orientation. These are the standard substrates for photonics.

The fabrication process flow is depicted in Fig. 3.2. An SOI wafer with 250 nm thick device layer and a buried oxide thickness of 3 μm is used as a substrate. A 70 nm thick SiO_2 hard mask is thermally grown at a temperature of 950°C. This way roughly 30 nm of Si are consumed, and the thickness of the remaining Si reduces to 220 nm. A negative-tone photo resist (PR) is used to define the waveguide pattern by using electron beam lithography. The PR pattern is transferred to the hard mask either by wet etching using buffered hydrofluoric acid, or by selective reactive ion etching in a CHF_3/O_2-plasma. Next, the 220 nm thick silicon layer is etched in a 40 % aqueous solution of

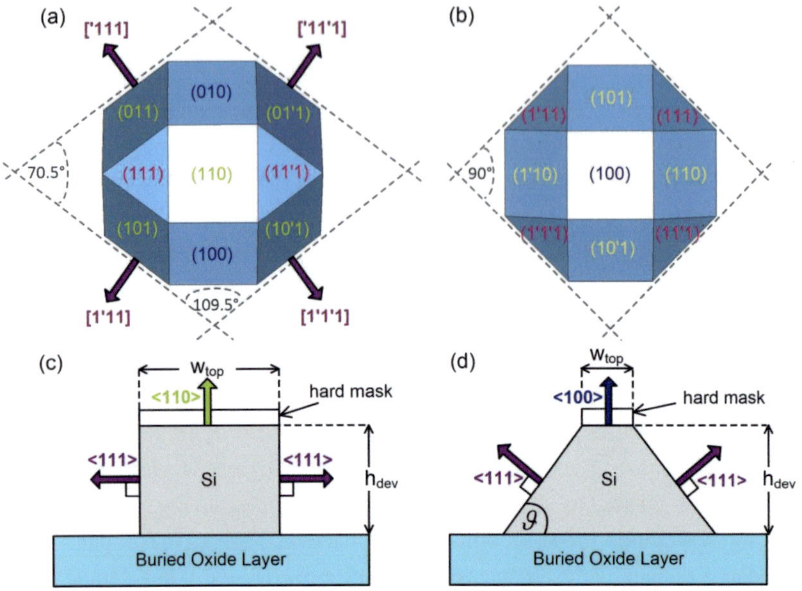

Fig. 3.1 Crystal planes in silicon. (a) (110)-SOI wafer. (b) (100)-SOI wafer. Anisotropically etched silicon waveguide with (c) rectangular cross-section on a (110)-SOI substrate, and with (d) trapezoidal cross-section on a (100)-SOI substrate. $\vartheta = 54.74°$.

KOH at a process temperature of 40°C, resulting in an etch rate of approximately 80 nm/min. A longer etch rate results in a stronger smoothening of the waveguide sidewalls, but also in a slight under-etching of the hardmask due to the small etch rate of the {111}-planes. After anisotropic wet etching, the SiO_2 hard mask is removed using hydrofluoric acid. Finally, an RCA clean [58] is performed to remove traces of organic and metallic contaminations.

Next, we will investigate the mode field profile of a trapezoidal waveguide and determine its regime of singlemode operation, followed by a characterization of the fabricated devices.

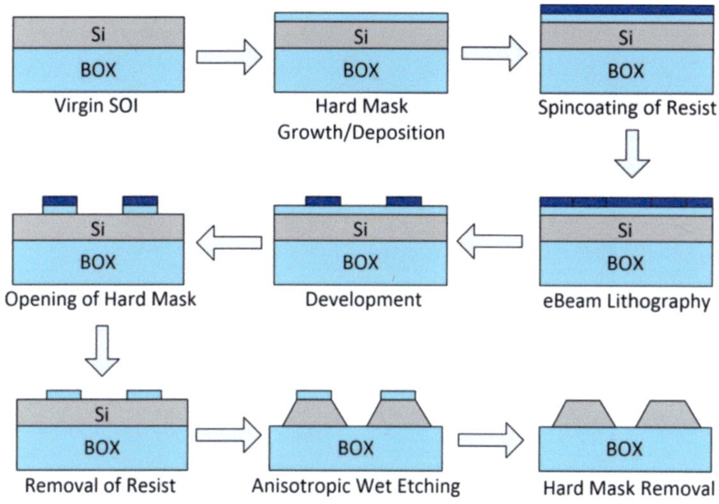

Fig. 3.2 Fabrication process flow.

3.2.1 Trapezoidal Waveguides

The optical mode fields of the first three fundamental modes of a trapezoidal WG with a height of 220 nm and a width $w_{\text{top}} = 400$ nm are depicted in Fig. 3.3. The first mode is a quasi-TE mode and the second mode is a quasi-TM mode. The third mode, a quasi-TE mode of second-order is only weakly confined at this waveguide width. Obviously, the waveguide is already close to being multimoded.The effective refractive indices of the first three modes of waveguides with trapezoidal and rectangular cross-section are depicted in Fig. 3.4 as a function of waveguide width. Up to a waveguide width of $w_{\text{top}} = 350$ nm the trapezoidal waveguide guides only the fundamental quasi-TE and quasi-TM mode. Between a width of $w_{\text{top}} = 400$ nm and $w_{\text{top}} = 600$ nm an anti-crossing of the second mode and of the third mode can be observed. In the anti-crossing region the notion of quasi-TE or quasi-TM modes does not hold any more for the second and third mode, because E_x- and E_y-components are comparable in magnitude. Mode coupling between these two modes becomes strong due to the similarities of the fields and of the respective propagation constant. For top widths $w_{\text{top}} > 600$ nm the second guided mode is a second-order quasi-TE mode, and the third guided mode is the quasi-TM mode. A waveguide with rectangular cross-section also features an anti-crossing, as depicted in Fig. 3.4(b). The coupling of these modes is stronger, but spectrally narrower than for a trapezoidal waveguide.

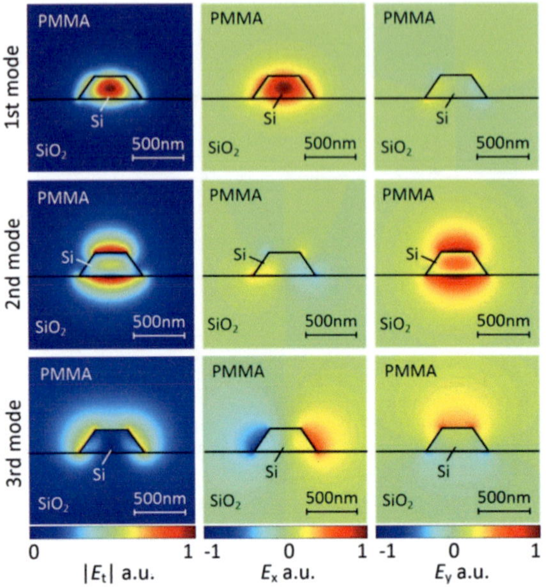

Fig. 3.3 Modes of a trapezoidal waveguide. The sidewalls are inclined at an angle of 54.74°. Depicted are the transversal electric field $|E_t|$, and the x- and y-component of the electric field of the first three modes of a waveguide with height of 220 nm and top width $w_{top} = 400$ nm. The modes can be denoted as first-order quasi-TE, first-order quasi-TM and second-order quasi-TE. PMMA is used as a cladding material in this simulation.

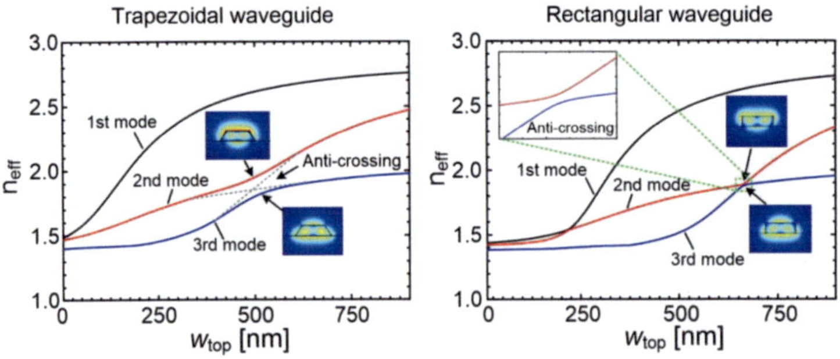

Fig. 3.4 Effective refractive index of a trapezoidal waveguide and of a rectangular waveguide as a function of waveguide width. An anti-crossing of the second and of the third mode can be observed. This anti-crossing is more pronounced in the case of the trapezoidal waveguide. In the anti-crossing region the modes become hybrid modes with mixed polarization.

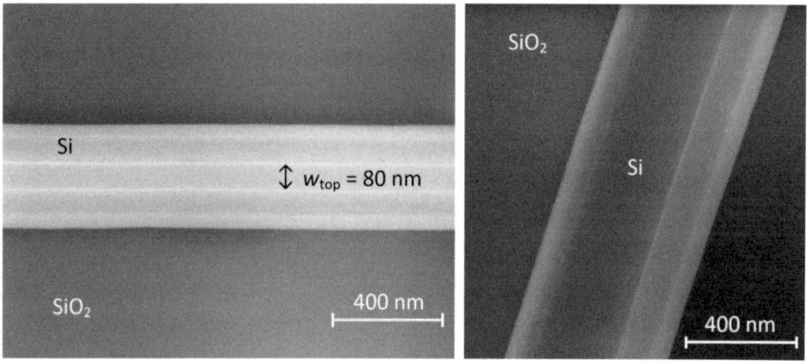

Fig. 3.5 SEM images of fabricated trapezoidal waveguides with smooth sidewalls.

We fabricate trapezoidal prototype waveguides of length between 1 mm and 10 mm and use trapezoidal grating couplers, as described in Section 3.2.3, for fiber-chip coupling. Two exemplary trapezoidal waveguides of width $w_{top} = 80$ nm and $w_{top} = 350$ nm are depicted in Fig. 3.5. The sidewalls are remarkably smooth.

We characterize waveguides with $w_{top} = 450$ nm using a tuneable laser source, a polarization controller and an optical spectrum analyzer. The recorded transmission spectra are displayed in Fig. 3.6(a). For waveguides of length of 1 mm, 4 mm, and 6 mm we observe almost identical insertion losses, indicating a low waveguide attenuation. However, we also observe an increase of non-periodic ripples in the measured spectra as the waveguide length increases. These ripples indicate the existence of random, discrete scattering centers along the waveguide, which are due to fabrication imperfections. These discrete waveguide defects prevent us from fabricating longer waveguides that would be required for accurately measuring the waveguide attenuation. At a waveguide length of 10 mm the ripples become severe and a 6 dB increase of insertion loss is observed. Nevertheless, we extract the waveguide attenuation from the measurement of the short waveguides by performing a linear regression of transmission vs. waveguide length. The extracted propagation losses in the wavelength range between 1520 nm and 1620 nm are depicted in Fig. 3.6(b). Averaging over the measurement wavelength range results in a loss of (0.1 ± 6.7)dB/cm when excluding the measurement of the 10 mm long waveguide and in a loss of (6.4 ± 6.3)dB/cm when including the 10 mm long waveguide to the linear regression. Obviously, the insuffucient waveguide length prohibits a precise measurement of the attenuation.

The origin of the ripples in the spectra was studied by performing an optical backscatter reflectometry (OBR) measurement, Fig. 3.6(c), with the commercial instrument OBR 4400, Luna Technologies. The OBR measurement confirms that discrete scatterers are the origin of the ripples in the spectrum. From this measurement, a loss of roughly 6 dB/cm can be extracted. Unfortunately, the fabrication of long, defect-free trapezoidal waveguides was not successful under the university cleanroom conditions. Thus these experiments can only give an indication to the actual potential of this technological approach.

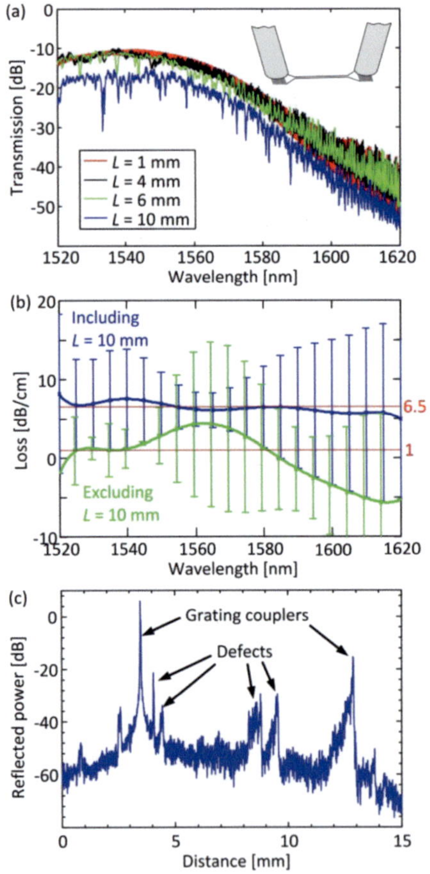

Fig. 3.6 Characterization of 450 nm wide trapezoidal waveguides (a) Transmission spectra of waveguides with different length L. (b) Fitted propagation loss. (c) Optical backscatter reflectometry of a trapezoidal waveguide.

3.2.2 Hybridization of Wet and Dry Etching

The utilization of anisotropic wet etching goes along with severe design restrictions. Depending on the specific integrated photonic circuitry it may be inevitable to combine anisotropic wet etching with reactive ion etching techniques, e. g., for realizing low-loss waveguide bends. In the case of a (100)-SOI substrate this requires a transition from a waveguide with trapezoidal cross-section to a waveguide with nearly rectangular cross-section. The simplest connection between these two geometries is an abrupt transition, as depicted in Fig. 3.7(a), where we couple a straight trapezoidal waveguide to a straight rectangular waveguide, followed by a 90° bend with 5 µm radius.

We investigate the insertion loss and the reflection of this transition as a function of lateral displacement between the waveguides. The simulated S-parameters for a transition between a trapezoidal waveguide with 300 nm top width and a rectangular strip waveguide with a width of 450 nm are depicted in Fig. 3.7(b). Without lateral displacement a loss of only 0.01 dB per tran-

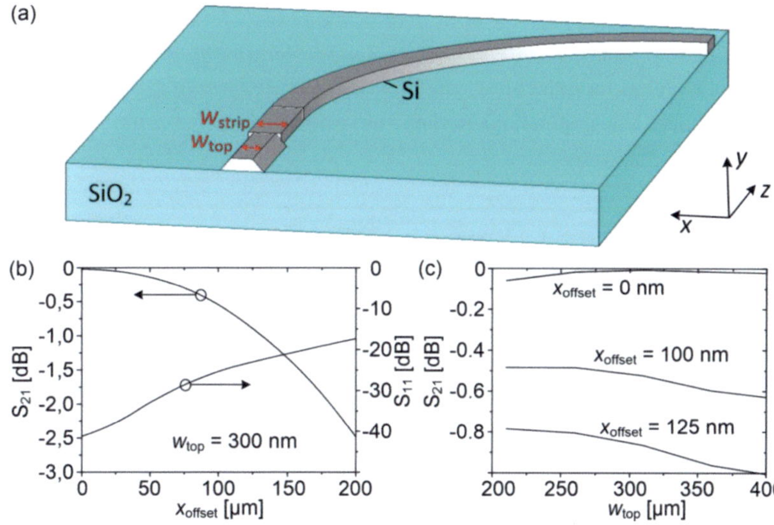

Fig. 3.7 Simulation of a transition between trapezoidal waveguide and rectangular waveguide. (a) Simulated structure. (b) Transmission (S_{21}) and reflection (S_{11}) of a transition between a trapezoidal waveguide of width w_{top} = 300 nm and a rectangular waveguide of 450 nm width as a function of lateral displacement in x-direction. (c) Performance of the transition for various widths of the trapezoidal waveguide. The width of the strip waveguide remains at w_{strip} = 450 nm.

sition is found with a low reflection factor of -40 dB. For a high misalignment of 100 nm the insertion loss increases to 0.52 dB and the reflection remains below -25 dB. Furthermore, the simulations show that such transitions are suitable for a multitude of trapezoidal waveguide widths, see Fig. 3.7(c).

3.2.3 Trapezoidal Grating Couplers

An essential building block of each photonic integrated circuit is a spot-size converter for coupling the integrated waveguides to standard singlemode fibers. In the following we will show that efficient grating couplers can be realized using anisotropic wet etching. We will begin with a simple uniform grating structure with exponential field profile. Afterwards, a non-uniform grating with Gaussian field profile will be presented.

3.2.3.1 Uniform Trapezoidal Grating

The design of a simple periodic grating is depicted in Fig. 3.8(a). For our simulations we choose a SOI thickness of 220 nm and a 2 µm thick buried oxide. A 1 µm thick layer of PMMA is used as a cladding material. In analogy to the design of gratings for plasma-etching-based fabrication [48], [51] we adapt the grating period Λ and the groove width w_{groove} of the grating, such that the first diffraction order is radiated at an angle of 10° with respect to the surface normal. Note that for a trapezoidal grating the groove width determines the etch depth h_{groove} of the groove

$$h_{\text{groove}} = \frac{w_{\text{groove}}}{2} \tan^{-1}(\vartheta) . \tag{3.3}$$

For highest coupling efficiency we find the optimized grating parameters $\Lambda = 635\ \text{nm}$, $w_{\text{groove}} = 210\ \text{nm}$, and a grating width of 15 µm. The simulated electric field profile of the grating is depicted in Fig. 3.8(b). The coupling efficiency of the grating η can be calculated as a product of the directionality D, describing the percentage of upwards radiated power, and the power coupling coefficient $\kappa^{(P)}$, see Eq. (B.20), derived from the overlap integral between fiber mode and radiated field of the grating,

$$\eta = D \cdot \kappa^{(P)} = \frac{P_{\text{output, fiber}}}{P_{\text{input, GC}}} . \tag{3.4}$$

For the uniform trapezoidal grating coupler we find a directionality $D = 0.58$ and a power coupling coefficient $\kappa^{(P)} = 0.85$, resulting in a power coupling efficiency η of 49.1 % (–3.1 dB). This value is slightly higher than the coupling efficiency of the corresponding uniform grating for dry etching, which is 44 % [48].

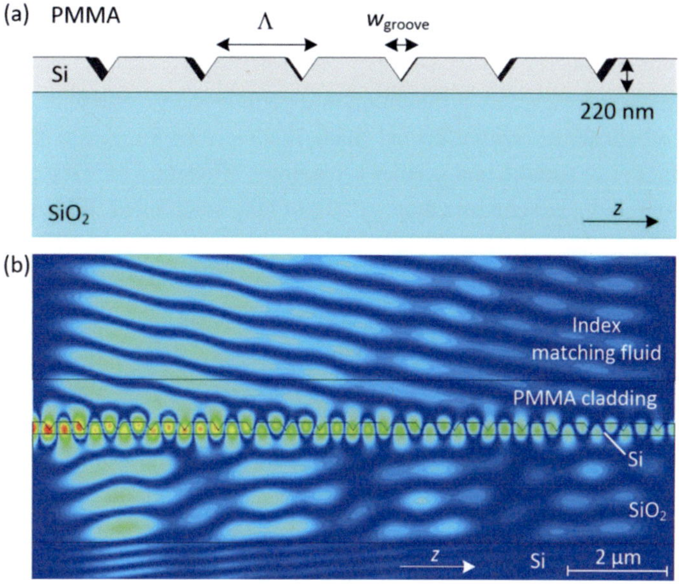

Fig. 3.8 Uniform trapezoidal grating coupler. (a) Schematic. (b) Simulated electric field propagation for an optimized grating with period $\Lambda = 635$ nm and groove width $w_{groove} = 210$ nm. The grating width is 15 µm and the first-order diffraction is radiated at an angle of 10°.

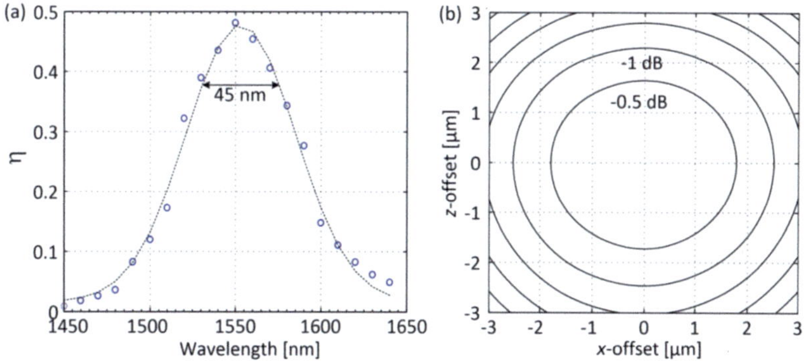

Fig. 3.9 Performance of a uniform trapezoidal grating. (a) Simulated coupling spectrum of a uniform grating coupler. The 1 dB-bandwidth is 45 nm. (b) Tolerance of the grating against lateral misalignment. Each line represents additional 0.5 dB coupling loss. A misalignment of ±2 µm results in a coupling penalty of less than 1 dB.

The simulated coupling spectrum of the grating is depicted in Fig. 3.9(a) and features a 1 dB-bandwidth of 45 nm. Similarly to dry etched grating couplers, the trapezoidal grating features high misalignment tolerances: As depicted in Fig. 3.9(b) a lateral fiber misalignment of ±2 μm results in less than 1 dB of additional coupling loss.

Sets of gratings with different grating periods and groove widths are fabricated on substrates with a buried oxide thickness of 3 μm. For this buried oxide thickness simulations predict a coupling efficiency of 44% (–3.6 dB). The two gratings are connected by a 200 μm long singlemode WG and by two 400 μm long linear tapers. Note that in contrast to fabrication of shallow etched gratings in dry etching technology, here WGs and grating couplers are etched in one etch step and require only one lithographic layer. A scanning electron micrograph of a fabricated trapezoidal grating is depicted in Fig. 3.10(a). The grating couplers are characterized at a coupling angle of 10° using a tuneable laser source and an optical spectrum analyzer. Two exemplary transmission spectra are depicted in Fig. 3.10(b). We find the best grating performance for a period of 605 nm and for a groove width of 180 nm. The measured coupling efficiency is –4.5 dB and the 1 dB-bandwidth is 32 nm. Note that no index matching fluid was used in this experiment. The application of an index matching fluid could result in an improvement of coupling efficiency by up to 0.5 dB.

Next the coupling efficiency of the grating is optimized by adapting the radiated field profile and by improving the directionality of the grating.

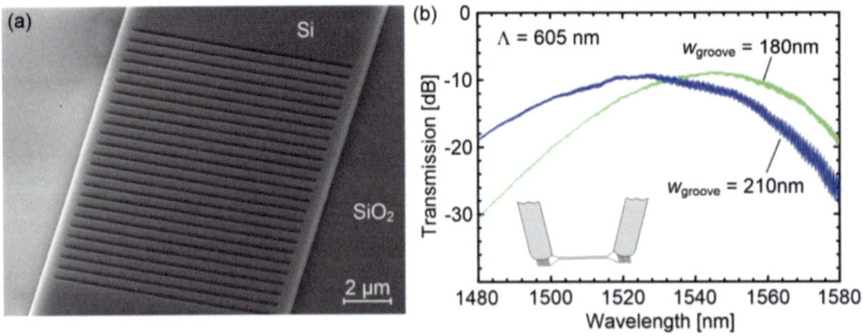

Fig. 3.10 Characterization of fabricated uniform grating couplers. (a) SEM image of a uniform trapezoidal grating. (b) Transmission spectra of two pairs of grating couplers. A 1 μm thick layer of PMMA is used as a cladding material. No index matching fluid is used to avoid reflection at the interfaces fiber-air and air-grating. We find a coupling efficiency of –4.5 dB.

3.2.3.2 Non-Uniform Trapezoidal Grating

According to Eq. (3.4) the coupling efficiency of a grating can be improved by maximizing both the power coupling coefficient $\kappa^{(P)}$ and the directionality D of the grating. In the following, we will first improve the power coupling coefficient by designing a non-uniform grating for matching the Gaussian field distribution of a fiber mode. In a later step we increase the directionality of the grating by using a metallic mirror beneath the buried oxide layer.

Non-uniform grating couplers with Gaussian beam profile have been realized on the silicon-on-insulator platform using a constant etch depth [59], [13] or by using a variable etch depth by exploiting the lag-effect [60] with power coupling coefficients as high as 97 % [59]. We follow the theoretical approach of references [61], [62] for designing a non-uniform grating that can be fabricated by anisotropic wet etching.

The decay of guided power $P(z)$ in a grating along the propagation direction z can be described by

$$\frac{\mathrm{d}P(z)}{\mathrm{d}z} = -\alpha_l(z)P(z) \tag{3.5}$$

where $\alpha_l(z)$ is the power leakage parameter of the grating. For a uniform grating the leakage parameter is independent of z, resulting in an exponential power decay in the grating

$$P(z) = P_0 \exp\left(-\alpha_l z\right). \tag{3.6}$$

In a grating the locally guided power $P(z)$ determines the light power that is radiated away from the grating per length, $B^2(z) = \alpha_l(z)P(z)$, therefore

$$\frac{\mathrm{d}P(z)}{\mathrm{d}z} = -B^2(z). \tag{3.7}$$

We integrate Eq. (3.7)

$$P(z) - P_0 = -\int_0^z B^2(z')\mathrm{d}z' \tag{3.8}$$

and substitute $P(z)$ in the RHS of Eq. (3.5) with Eq. (3.8), and in the LHS with Eq. (3.7). Consequently, we find an expression for the leakage parameter,

$$\alpha_l(z) = \frac{B^2(z)}{P_0 - \int_0^z B^2(z')\mathrm{d}z'}, \qquad P_0 = \int_0^\infty B^2(z')\mathrm{d}z'. \tag{3.9}$$

Using Eq. (3.9) we can calculate the power leakage parameter function $\alpha_l(z)$ that results in the emission of a certain radiated power $B^2(z)$ per length. For an efficient fiber-chip coupling the radiated power per length should be a Gaussian power distribution per length $B^2(z) = G(z)$, with Gaussian beam waist w_0 that is matched to the mode of the fiber. We define $G(z)$ as

$$G(z) = C_G \exp\left[-\frac{2(z - z_0)^2}{w_0^{\ 2}}\right], \qquad (3.10)$$

where C_G is a normalization factor of dimension W/m. For coupling to a standard singlemode fiber (SSMF) the mode field diameter $2w_0$ is set to 10.4 µm.

The calculated power leakage parameter function $\alpha_l(z)$ is plotted in Fig. 3.11(a) for a matching the mode profile of a SSMF. For optimizing the grating, we first determine the power leakage parameter of a single grating groove by simulation. Fig. 3.11(b) displays the simulated leakage parameters as a function of groove width. Next, to achieve first-order diffraction of the grating at an angle of 10° with respect to the surface normal, the grating period Λ has to be adapted for each groove width as depicted in Fig. 3.11(b). A schematic of a non-uniform grating that has been designed according to Section 3.2.3.2 and a simulation of its transversal electric field are depicted in Fig. 3.12(a,b). The radiated beam features the targeted Gaussian shape with a MFD = 10 µm and has a high power coupling coefficient $\kappa^{(P)} = 0.98$ to the mode of a singlemode fiber. Also the directionality of the grating improved to a value of $D = 0.61$, resulting in a coupling efficiency of 60% (–2.2 dB). The simulated coupling spectrum of a grating coupler is displayed in Fig. 3.13(a) and has a 1 dB-bandwidth of 48 nm. The non-uniform grating features high misalignment tolerances as depicted in Fig. 3.13(b), similarly to the uniform grating.

An SEM image of a fabricated non-uniform grating is shown in Fig. 3.14(a). It was fabricated on a SOI substrate with 3 µm buried oxide. For this oxide thickness simulations predict coupling losses of 2.7 dB. A measured transmission spectrum of a pair of couplers is depicted in Fig. 3.14(b). No index matching fluid was used in this measurement. The grating has a measured coupling efficiency of -5.7 dB and a 1 dB-bandwidth of 32 nm. In contrast to expectations the measured coupling efficiency is lower than the efficiency of a uniform trapezoidal waveguide which is probably due to deviations of the fabricated groove widths from the targeted ones. Furthermore a difference in grating resonance of the two involved non-uniform gratings could have resulted in the decrease in coupling efficiency.

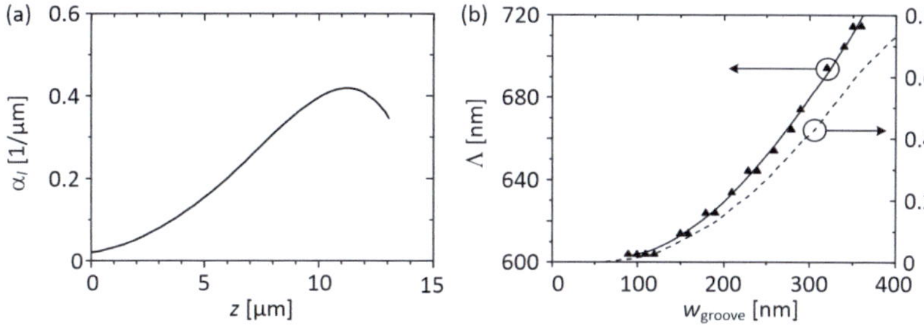

Fig. 3.11 Design of a non-uniform GC with Gaussian beam profile. (a) Power leakage parameter distribution $\alpha_l(z)$ for matching a Gaussian beam with mode field diameter $2w_0 = 10.4\,\mu\text{m}$. (b) Grating period Λ for radiation at an angle of 10° to the surface normal as a function of groove width w_{groove} together with the corresponding leakage parameter (RHS axis).

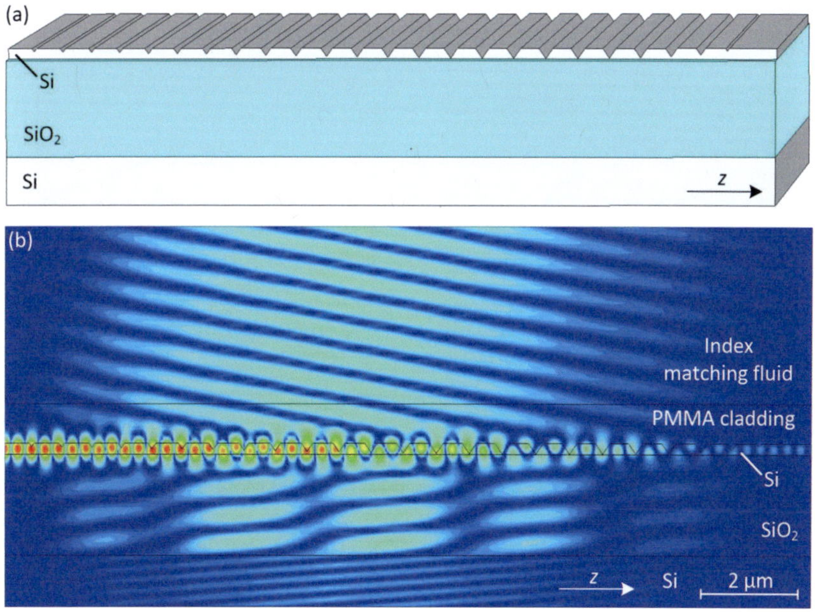

Fig. 3.12 Non-uniform trapezoidal grating. (a) Schematic (b) Simulation of the transverse electric field magnitude along the z-direction.

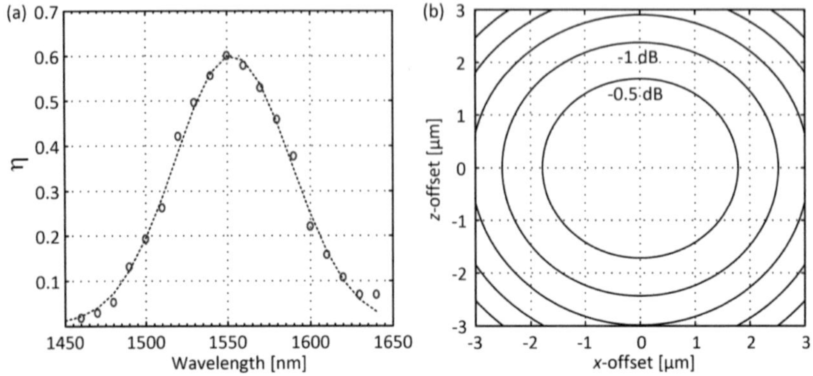

Fig. 3.13 Performance of a non-uniform trapezoidal grating. (a) Simulated coupling spectrum of a non-uniform grating coupler. The 1 dB-bandwidth is 48 nm. (b) Tolerance of the grating against lateral misalignment. Each line represents additional 0.5 dB coupling loss. A misalignment of ±2 µm results in a coupling penalty of less than 1 dB.

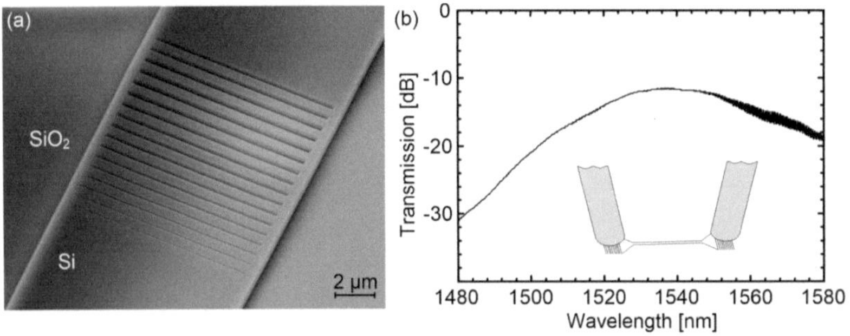

Fig. 3.14 Characterization of a fabricated non-uniform grating coupler. (a) SEM image of a non-uniform trapezoidal grating. (b) Transmission spectra of a pair of GC. A 1 µm thick layer of PMMA is used as a cladding material. No index matching fluid is used that would avoid reflection at the interface fiber-air. We find a coupling efficiency of –5.7 dB.

3.2.3.3 Non-Uniform Grating with Metallic Bottom Reflector

The coupling efficiency of a non-uniform grating coupler can be further enhanced by increasing directionality. It has been shown that a metallic reflector beneath the buried oxide (BOX) layer can be used to strongly enhance directionality [52]. This way, grating couplers with coupling efficiency of –0.6 dB have been reported [13]. Key for enhanced directionality is the constructive superposition of the downwards diffracted wave and the reflected wave. Thus the thickness of the buried oxide has to be properly chosen. A simulation of the coupling efficiency of a non-uniform grating with metallic bottom reflector is depicted in Fig. 3.15 as a function of buried oxide thickness. An optimum coupling efficiency of 93% (-0.32 dB) is found for a BOX thickness of 2 μm. The corresponding simulation of the grating is depicted in Fig. 3.16. The performance of these optimized trapezoidal grating couplers is comparable to the most advanced gratings fabricated by dry etching, where a coupling efficiency of –0.33 dB and a 1 dB-bandwidth of 44 nm are expected from simulations.

We thus conclude that anisotropic wet etching in combination with electron beam lithography is a promising technology for the fabrication of low loss photonic integrated circuits, where waveguides and gratings can be fabricated in a single etch step. Depending on the specific application, wet etching can compete with dry etching, and in addition enables novel waveguide geometries that cannot be realized by state-of-the-art dry etching technology.

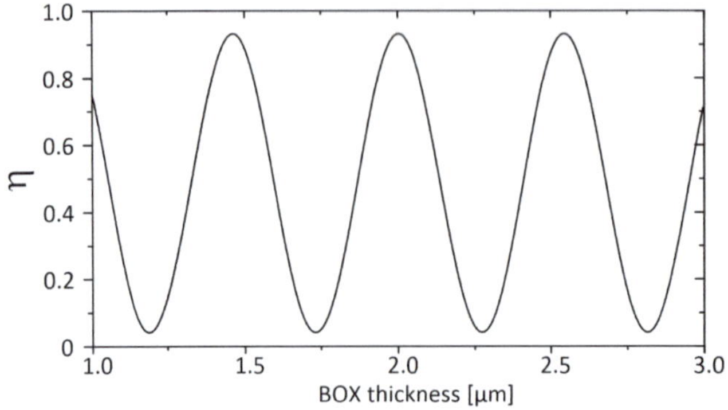

Fig. 3.15 Coupling efficiency of a non-uniform trapezoidal grating with bottom reflector as a function of the buried oxide (BOX) thickness. The coupling angle is 10° with respect to the surface normal. Operating wavelength $\lambda = 1550$ nm.

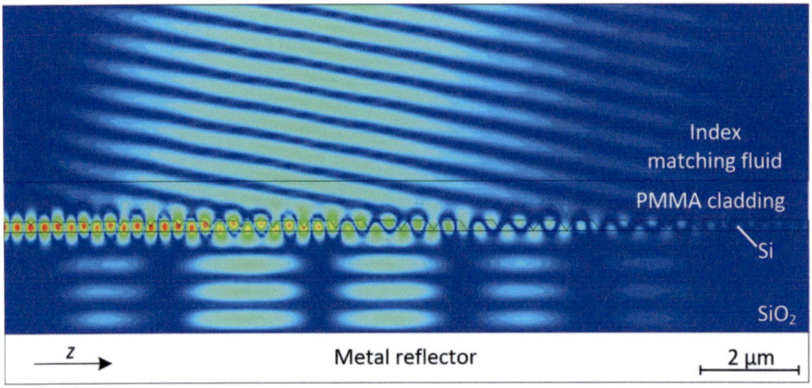

Fig. 3.16 Simulated transversal electric field magnitude of a non-uniform trapezoidal grating with metallic bottom reflector. The thickness of the buried oxide is 2 µm. We find $\kappa^{(P)} = 0.98$, $D = 0.95$, and a coupling efficiency of $\eta = 0.93$ (-0.32 dB). The coupling angle is 10° with respect to the surface normal. Operating wavelength $\lambda = 1550$ nm.

3.3 Anisotropic Wet Etching and Optical Contact Lithography

Finally, it will be shown that smooth nanowire waveguides can be fabricated by standard contact UV lithography and anisotropic wet etching. This is achieved by utilizing a special staggered mask geometry.

The smallest features that can be fabricated by optical contact lithography are in the order of 1.5 µm. For fabricating sub-micrometer structures, masks composed of staggered rectangular elements of 2 µm width and 5 µm length are used. The offset between the single elements has been varied to achieve different overlaps of the elements, see Fig. 3.17(a). The overlap area of the staggered elements defines the resulting width of the silicon waveguides, because anisotropic wet etching under-etches convex corners, until a {111}-plane is formed.

The staggered mask is applied to an SOI-sample with (110)-crystal orientation, and to a sample with (100)-crystal orientation. The sample with (110)-orientation has a silicon device layer thickness of 1.6 µm. At the time of this experiment (110)-SOI wafers with proper device layer thickness were not available. Still, such wafers are good enough to demonstrate the principle. The resulting structure of the (110)-sample after etching is depicted in Fig. 3.17(c). The staggered mask results in a meander-shaped SiO_2 hard mask. After 1 h of etching the convex corners of the hard mask are completely under-etched and only the overlapping area remains. The cross-section of the resulting rectangular structure after hard mask removal is depicted in Fig. 3.17(d). The structure is only 200 nm wide and 1.6 µm high, which corresponds to a high aspect ratio of 8:1. Next, the staggered mask is applied to a (100)-SOI sample with 220 nm thick silicon device layer. The resulting trapezoidal nanowire is depicted in Fig. 3.17(e). The sidewall of this waveguide is scanned using atomic force microscopy, Fig. 3.17(f). A small root mean square surface roughness of rms = 1…3 nm is found.

Therefore, ultra-smooth nanowire strip waveguides can be fabricated with nanoscale precision by using conventional optical lithography in combination with a special staggered mask scheme and anisotropic wet etching. The fabricated structures are far smaller than the resolution limit of the underlying lithographical system.

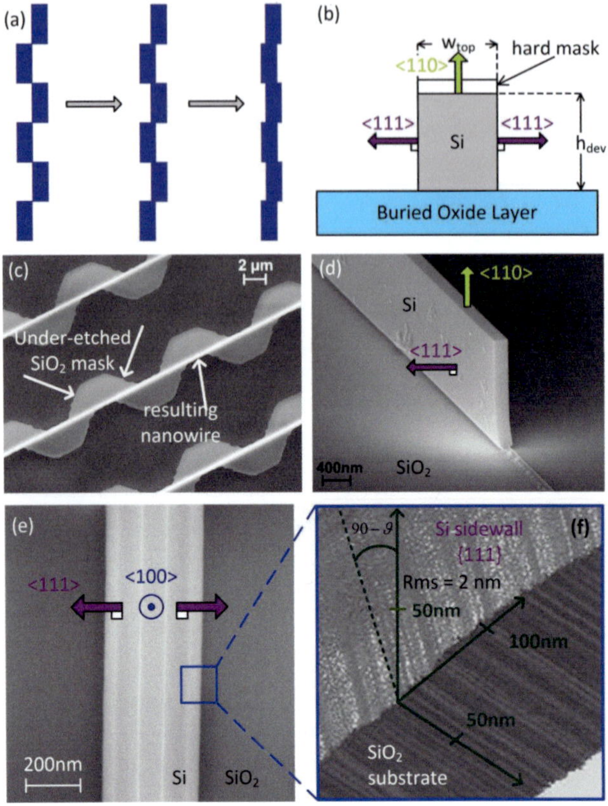

Fig. 3.17 Waveguide fabrication using staggered masks. (a) Cross-section of an anisotropically etched strip waveguide in the case of a (110)-SOI wafer. (b) Mask scheme using staggered rectangular elements. The area where the staggered micrometer-scale elements overlap defines the resulting nanowire waveguide width. (c) SEM image of sub-micrometer wide strip waveguides. The thin staggered oxide hard mask (rounded due to the resolution limit) was left on top of the waveguides to illustrate the under-etching of the convex corners. (d) Cross-section of a 200 nm wide and 1.6 µm high (multimoded) waveguide with vertical sidewalls after hard mask removal. (e) Staggered masks on a (100)-SOI wafer. SEM image of a smooth trapezoidal waveguide. Top width 90 nm, height 220 nm, bottom width 370 nm. (c) AFM image of the sidewall of a trapezoidal waveguide. The measured rms roughness varies between 1..3 nm. Modified after [C5]. Copyright © 2011 IEEE

3.4 Summary

Smooth nanowire waveguides and efficient grating couplers can be fabricated by anisotropic wet etching in an aqueous potassium hydroxide solution. First prototypes with trapezoidal cross-section featured propagation losses around 6 dB/cm, mainly caused by discrete waveguide defects of these prototypes. We expect that much lower losses can be achieved when working under better controlled processing conditions. The fabricated grating couplers have a coupling loss of 4.5 dB, while simulation show that losses of these couplers could be as low as 0.3 dB when using a non-uniform grating with metallic bottom reflector. Furthermore, it has been shown that smooth nanowires can be fabricated by low-resolution optical contact lithography when exploiting a special mask layout that is composed of staggered elements in combination with anisotropic wet etching.

4 Horizontal Fiber-Chip Coupling using Angled Fibers

4.1 State of the Art

Efficient coupling between an integrated waveguide and an external glass fiber is a key requirement for any photonic integrated circuit platform. In silicon photonics mainly two approaches have been used for efficient fiber-chip coupling – inverted taper spot-size converters, see Section 2.3.1, and grating couplers, see Section 2.3.2.

Inverted tapers enable efficient in-plane fiber-chip coupling and a large optical bandwidth, but require cleaved fiber facets, which makes wafer-scale testing of devices challenging. An elegant SSMF coupling solution with passive fiber alignment has been demonstrated by etching an alignment V-groove into the silicon substrate of an SOI sample [10]. The remaining mode size mismatch between the inverted taper in the chip and the SSMF results in moderate coupling losses of 3.5 dB [10]. High coupling efficiencies usually require the additional use of tapered and/or lensed fibers that result in additional product cost. However, losses down to 1 dB [11] have been reported. Recent developments in three-dimensional direct laser writing using two-photon polymerization offer new possibilities for fiber-chip coupling with inverted tapers: A 3D polymer waveguide taper that surrounds the inverted silicon taper can be used to increase coupling efficiency between SSMF and inverted silicon taper. This directly-written 3D polymer waveguide is a special form of a so-called photonic wire bond [12], where chip-to-chip interconnects with coupling losses of 1.6 dB could be demonstrated.

Grating couplers, in contrast to inverted tapers, enable wafer-scale testing of devices by close-to-vertical fiber-chip coupling. Experimental chip-to-SSMF coupling efficiencies of –0.6 dB have been shown [13]. Furthermore, the misalignment tolerances between grating and fiber allow for a lateral displacement in the order of 2 µm. Disadvantages of this approach are the requirement of nearly vertical chip-coupling, which is impractical for packaging, and the limited optical bandwidth of the grating.

In this chapter a workaround for the vertical assembly of fiber and chip is studied, making use of total internal reflection at the polished angled facet of a

SSMF. This angled fiber concept was independently developed at the Tyndall National Institute and published in 2012 [14].

4.2 Angled Fiber Concept

This fiber coupling concept combines the advantages of grating couplers, see Section 2.3.2, such as wafer-scale testing compatibility, and relaxed alignment tolerances, with the benefits of a horizontal fiber arrangement by making use of polished angled fibers. A schematic of the in-plane coupling arrangement is depicted in Fig. 4.1(a). The facet of a SSMF is polished at an angle of 40°. This way the incident fiber beam is reflected at the fiber facet by total internal reflection and the beam exits the fiber through the fiber cladding at an angle of 10° with respect to the cladding normal. The out-coupling direction thus matches the coupling angle of the grating coupler that is depicted in Fig. 2.9 when placing the fiber horizontally over the grating. Detrimental reflections at the boundaries fiber-air and air-chip are avoided by bringing the "angled fiber" in contact with the chip: The refractive index of the glass fiber and of the chip cladding (either SiO_2 or PMMA) are very similar, thus "index matching" is achieved.

The electric field propagation in an angled fiber is simulated using the commercial simulation tool CST Microwave Studio and is depicted in Fig. 4.1(b). As expected, total internal reflection is observed at the fiber facet. After reflection the beam traverses the fiber cladding. The unguided beam can be approximated by a Gaussian beam. During propagation the mode field

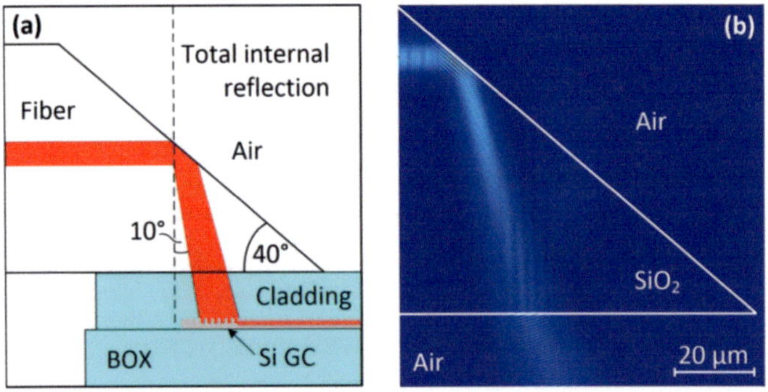

Fig. 4.1 Angled fiber concept (a) and simulated electric field propagation (b). The fiber facet is polished at an angle of 40°. The incident beam is thus totally reflected at the fiber facet and exits the fiber through the fiber cladding.

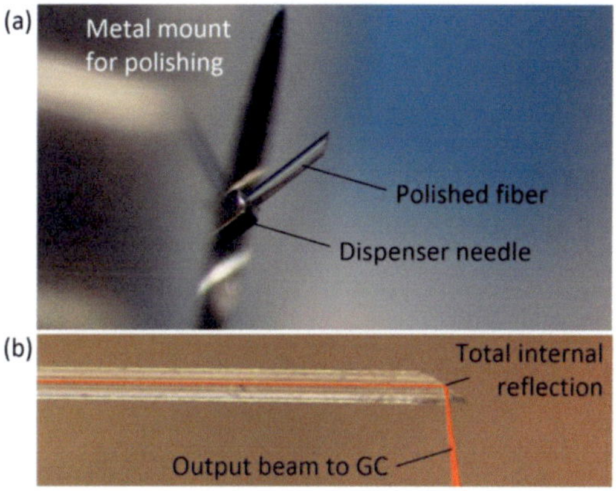

Fig. 4.2 Fabricated angled fiber. (a) Angled fiber in metal mount after polishing. A dispenser needle is used to hold the fiber in the metal mount. (b) Image of a fiber after polishing. The beam propagation is sketched in red color for illustration purposes. Pictures courtesy of Sven Schüle (KIT).

diameter of the Gaussian beam increases, and the initially planar phase fronts become curved phase fronts. The output field profile will be further investigated in Section 4.3 along with designing a grating coupler for an optimum coupling to the given beam shape.

The angled fiber is fabricated by polishing. The desired polishing angle is determined by an angled metal mount. The metal mount possesses a drilled hole for positioning a fiber. A dispenser needle is used for mechanical stabilization of the fiber. The assembly is glued together and the fiber is embedded in epoxy for polishing – this also prevents breaking of the fiber edges. After polishing, acetone is used remove the epoxy. A photo of a polished fiber inside its metal mount is depicted in Fig. 4.2(a). An angled fiber along with a schematic of the beam propagation is depicted in Fig. 4.2(b).

4.3 Beam Analysis and Grating Coupler Design

Next, the radiated beam profile of an angled fiber is analysed. The mode field diameter of the out-coupled beam is determined and the quality of the polished fiber facet is checked. The predicted beam profile is extracted from the simulation in Fig. 4.1(a) and depicted in Fig. 4.3(a). The beam diameter (mode field diameter – $\mathrm{MFD} = 2w_0$) increased from 10.4 µm to roughly 13.5 µm when traversing a length of ~60 µm in the cladding. In addition, the diameter of the out-coupled beam of a fabricated fiber is measured with a microscope, see

Fig. 4.3(b). A diameter of roughly 13 µm is measured. The beam diameter is slightly smaller in x-direction than in z-direction, because the fiber cladding acts as a cylindrical lens. No beam distortions from the polished fiber facet are observed in Fig. 4.3(b). It can hence be concluded that the quality of the polished facet suffices.

Knowing the radiated beam profile of the angled fiber the grating coupler can be adapted to match this beam profile. The GC needs to radiate a two-dimensional Gaussian beam profile with a beam diameter of 13.5 µm. In the following the curvature of the phase fronts of the Gaussian beam is neglected. A non-uniform trapezoidal grating coupler design is chosen for achieving a Gaussian beam profile, as described to Section 0. A metallic bottom reflector is used to increase directionality of the grating. A BOX thickness of 2 µm enables constructive interference of downwards radiated field and upwards reflected field. The optimized width of the grating is 16 µm. The groove widths and the corresponding periods of the non-uniform grating are calculated using the analytic approach of Section 3.2.3.2 and are listed in Table 1. A simulation of the radiated beam is depicted in Fig. 4.4. For coupling this grating to an angled fiber, a power coupling coefficient of $\kappa^{(P)} = 0.844$ and a directionality of $D = 0.949$ are expected. This results in a power coupling efficiency of $\eta = 0.801$ (–0.96 dB).

So far, the GC beam was only adapted to the spatial intensity of the fiber beam. The coupling efficiency could be further enhanced by adjusting amplitude and phase of the grating beam to match the curved phase front of the angled fiber beam.

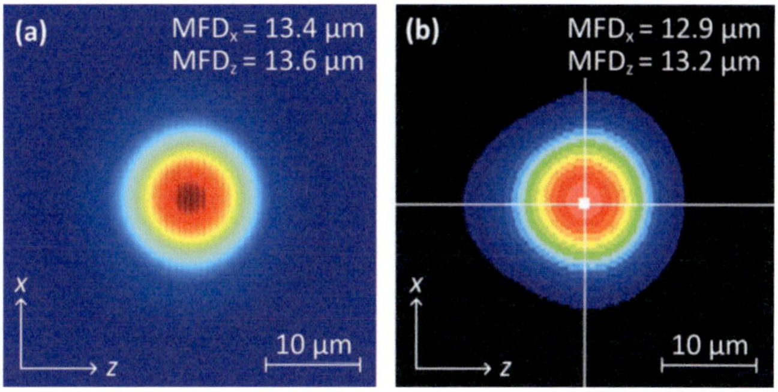

Fig. 4.3 Simulated and measured output beam profile of the angled fiber. A MFD of roughly 13.5 µm is extracted from the simulation in Fig. 4.1(b). The MFD in x-direction is slightly smaller since the fiber cladding acts as a cylindrical lens. (b) The mode is measured with an optical microscope. The measured MFD of 13 µm is slightly smaller than predicted by simulations.

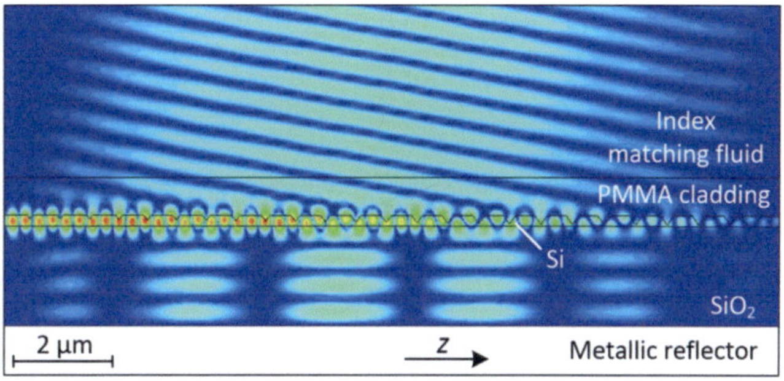

Fig. 4.4 A non-uniform grating coupler is adapted to match a Gaussian beam with mode field diameter $2w_0 = 13.5\ \mu\text{m}$. The coupler is designed for wet-chemical fabrication. A metallic bottom reflector is used to increase the directionality of the coupler. The predicted power coupling efficiency is $\eta = 0.801$.

Table 1 – Calculated set of groove widths and periods of a grating optimized for matching the mode of an angled fiber. The grating is depicted in Fig. 4.4.

Index	Groove width [µm]	Period [µm]
1	0.100	0.604
2	0.106	0.605
3	0.113	0.605
4	0.121	0.607
5	0.129	0.608
6	0.137	0.610
7	0.146	0.612
8	0.156	0.614
9	0.166	0.617
10	0.176	0.620
11	0.186	0.624
12	0.196	0.627
13	0.205	0.631
14	0.214	0.635
15	0.223	0.638
16	0.236	0.642
17	0.237	0.645
18	0.243	0.648
19	0.248	0.650
20	0.252	0.652
21	0.255	0.654
22	0.256	0.655
23	0.256	0.654
24	0.254	0.653

4.4 Measured Coupling Efficiency

The coupling efficiency between angled fiber and a 10 μm wide uniform grating is measured and compared to the coupling efficiency between cleaved fiber and grating. The measured transmission spectra are depicted in Fig. 4.5. A higher coupling efficiency is expected when using cleaved fibers, since this particular grating coupler is better matched to the mode of a cleaved fiber. However, the transmission through the cleaved fibers (red curve) is found to be 0.2 dB worse than transmission through angled fibers. This can be understood in the following way: When using cleaved fibers reflection occurs between fiber facet and air and between air and grating coupler cladding. This is not the case when using angled fibers. The cladding of the angled fiber is in contact with the cladding of the grating and acts like an index-matching material. However, when using cleaved fibers in combination with an index matching fluid (blue curve) an even higher transmission can be observed. Thus, coupling with angled fibers results in a coupling penalty of 0.45 dB per coupler as compared to regular cleaved fibers with index matching fluid.

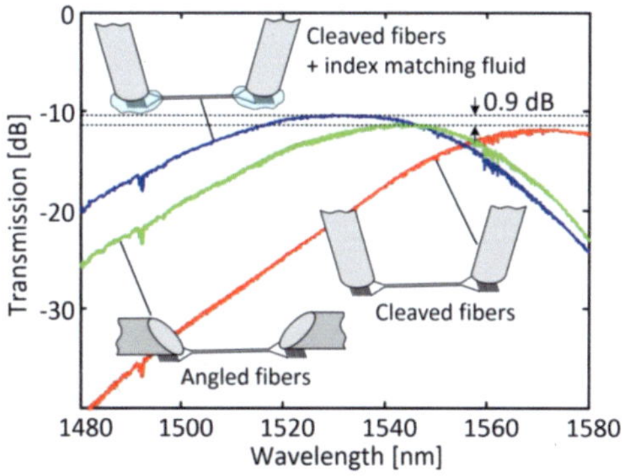

Fig. 4.5 Comparison between angled fiber (green) and cleaved fiber (red) when using non-optimized uniform grating couplers of 10 μm width. When using a cleaved fiber in combination with an index matching fluid (blue), reflections at the boundary fiber-air are reduced, and the coupling efficiency per fiber is 0.45 dB higher as compared to the angled fiber. This penalty is expected since the GC was not designed for operation with the angled fiber. Furthermore we observe a blue-shift of the grating resonance when using angled fibers or an index matching fluid, since the modified GC cladding modified the effective refractive index of the grating.

Furthermore, a blue-shift of the resonance wavelength can be observed when switching from cleaved fibers to angled fibers. A similar tendency is found when using cleaved fibers with index matching fluid. The reason for this blue-shift is a change of the propagation constant of the grating when adding an index matching fluid. A cladding thickness of 1 µm is thus not sufficient for decoupling grating and environment.

4.5 Misalignment Tolerance

Finally, the tolerance of the angled fiber with respect to translational and rotational misalignment is investigated. The coupling power penalty as a function of the vertical separation between fiber and grating is depicted in Fig. 4.6(a) for two different grating coupler widths. For both gratings the penalty is less than 1 dB when keeping the distance between fiber and grating below 70 µm. The lateral misalignment tolerance is depicted in Fig. 4.6(b). The coupling penalty is less than 1 dB for a lateral misalignment of up to 2.5 µm. Therefore coupling between angled fiber and grating is even more robust against lateral misalignment as compared to a cleaved SSMF, see Fig. 3.13, that is coupled out-of-plane.

In contrast to coupling with regular cleaved fibers, an angled fiber has a rotational degree of freedom that needs to be adjusted. The sensitivity of the coupling scheme is studied with respect to rotational misalignment,

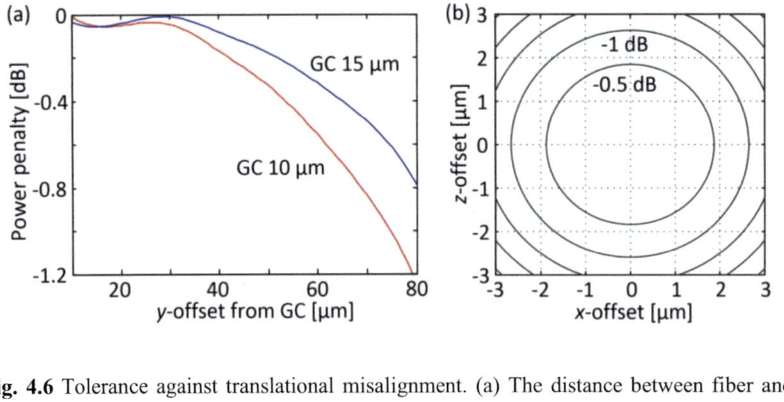

Fig. 4.6 Tolerance against translational misalignment. (a) The distance between fiber and GC is varied. A grating of 10 µm width and a grating of 15 µm width are investigated. (b) The fiber is misaligned in *x*- and *z*-direction. The curves represent the coupling penalty in steps of 0.5 dB. The investigated grating is the grating coupler from Fig. 4.4, which is matched to the beam of an angled fiber. A misalignment of ±2.5 µm results in less than 1 dB of additional loss.

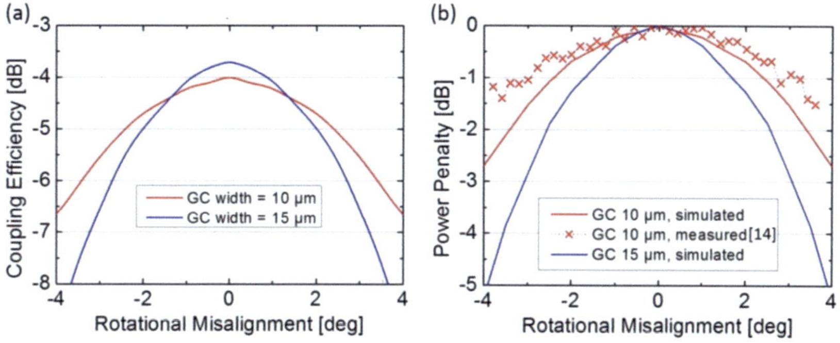

Fig. 4.7 Rotational misalignment of the angled fiber. A grating coupler of 10 µm width and of 15 µm width is investigated. (a) Power coupling efficiency. A non-optimized uniform grating coupler without bottom reflector is used. (b) Coupling penalty.

see Fig. 4.7. For a grating width of 15 µm, nearly matched to the beam of an angled fiber, a coupling penalty due to rotational misalignment of less than 1 dB is found for a rotational offset below 1.7°. Interestingly, the alignment tolerance is higher (~2.3°) when using a narrower grating of 10 µm width. However, the total coupling efficiency of the narrow grating is worse than the one of the wide grating if the fiber is correctly aligned. The simulated rotational misalignment is in good agreement with experimental results that are presented in Ref. [14].

4.6 Summary

In this chapter a novel in-plane fiber-chip coupling method was demonstrated using a specially polished fiber and a surface emitting grating coupler. When using matched gratings, a coupling loss below 1 dB is expected. Due to the increased diameter of the output beam of the fiber, alignment tolerances are slightly higher as compared to vertical coupling with cleaved fibers.

5 Silicon-Organic Hybrid (SOH) Modulator

In this chapter, a short overview on the state of the art of silicon modulators is given first, followed by a detailed description of the investigated SOH modulator. The performance of the individual passive building blocks of the modulator is studied. Various EO organic materials for modulators are investigated and compared. Furthermore, the RF design of the device, its static and dynamic modulation characteristics, and the modulator's energy consumption are described in detail.

5.1 State of the Art

Within the last decade silicon became a promising material for high-density integrated photonic circuits. Interest focused especially on the realization of optical interconnects for data centers, metro networks, or fiber-to-the-home customers, leveraging mature high-yield CMOS processing and offering the potential of photonic-electronic co-integration on large-area silicon wafers. Current optical interconnects in data centers are based on directly modulated VCSELs. By parallelization of up to 12 VCSELs, each operating at 12.5 Gbit/s, aggregate data rates of up to 150 Gbit/s [63] are currently achieved. Although individual VCSEL dies were shown to operate at up to 60 Gbit/s [64], reliability [65] of such devices makes their deployment challenging even at an intermediate data rate of 25 Gbit/s. In order to increase data rates further, complex modulation formats are likely to be used in future interconnects. Complex modulation formats are not possible with intensity modulated VCSELs, or using electro-absorption-based devices. Therefore the development of high-bandwidth phase and amplitude modulators on the silicon platform is of particular importance. In addition to high data rates, silicon modulators need to provide ultra-low energy consumption of only a few fJ per bit [66] to keep the power dissipation at a reasonable level.

5.1.1 All-Silicon Modulators

Commercial amplitude modulators on material platforms like $LiNbO_3$, InP or GaAs rely on the Pockels effect. However, since second-order nonlinearities are absent in bulk silicon due to inversion symmetry of the crystal lattice, all-silicon modulators have to rely on the plasma effect instead. So far, silicon modulators have been reported that are based on free-carrier depletion or injection in *p-n*, *p-i-n*, or metal-oxide-semiconductor (MOS) structures [17]. This leads to various trade-offs when realizing fast and energy-efficient devices with small footprint: While carrier injection in forward-biased *p-i-n*-structures enables compact modulators with voltage-length products as low as $U_{\pi}L = 0.36\,\text{Vmm}$ [18], the free-carrier lifetime currently limits the modulation speed to 25 Gbit/s and requires strong pre-emphasis of the drive signal [19]. Moreover, the energy efficiency of these devices is intrinsically limited to the pJ/bit-range due to the injection current flowing permanently through the diode. In contrast, carrier depletion in reverse-biased *p-n* junctions enables negligible bias currents and symbol rates of up to 50 GBd [20], but typical voltage-length products are larger than 10 Vmm and thus much larger than those of injection-type devices. For a silicon-based Mach-Zehnder modulator (MZM), the lowest reported energy consumption amounts to 200 fJ/bit, achieved in a 5 mm long depletion-type device [23]. Modulation energies and device footprint can be significantly reduced by using ring resonators, micro-disc resonators, or photonic-crystal waveguides. For resonant silicon-based modulators, the lowest energy consumption reported to date amounts to 3 fJ/bit and has been achieved with a microdisc device [24] operated at a data rate of 12.5 Gbit/s. However, the optical bandwidth of resonant devices is inherently limited, and resonance wavelengths are often subject to strong temperature-induced drifts. Apart from the mentioned intrinsic limitations of silicon, its basic feasibility for the realization of integrated IQ-modulators has been demonstrated. QPSK modulation at a symbol rate of 10 GBd has been demonstrated using carrier-depletion ring resonator modulators [67], while QPSK and 16QAM modulation at a symbol rate of 28 GBd was demonstrated using a carrier-depletion MZM [21],[22].

5.1.2 SOH Modulators

A fundamentally different approach for building modulators on the silicon platform is the so-called "silicon-organic hybrid (SOH)" approach [26], [46]. Here, nonlinear optical effects of second order (Pockels effect) are exploited through interaction of a guided light mode in silicon with an organic $\chi^{(2)}$-cladding material. In principle, the SOH approach can simultaneously meet all modulator requirements such as low drive voltage, high optical and electrical bandwidth, compact size, and low energy consumption [26], [29].

The operation principle of a slotted SOH modulator is described in Section 2.4. The silicon slot is infiltrated by an EO organic material. The most frequently employed class of second-order nonlinear organic claddings for

SOH integration are nonlinear optic molecules, so-called chromophores as dopants in polymeric host-matrices [30]–[35]. Prominent examples are YLD124 in PMMA, M1, M3 (commercial products of GigOptix with unknown chemical composition), and AJLZ53 in PMMA. Compared to recently introduced approaches [68], these nonlinear materials show moderately high nonlinear coefficients, e. g., poling of µm-thick films of 25wt.% YLD124 in PMMA resulted in $r_{33} = 118$ pm/V [69]. Within the SOH approach, this would result in a π-voltage $U_\pi = 0.75$ V for a 1 mm long MZM with a moderate slot width of 160 nm and a rail width of 210 nm. The expected energy consumption would be 56 fJ/bit (modulator represented by a lumped capacitor only, see Eq. (5.10)). However, in all cases discussed in literature, much lower EO coefficients were found in SOH devices compared to the parallel-plate poled bulk references. Table 2 gives an overview on a variety of EO organic claddings and their actual EO coefficients in SOH devices. In-device EO coefficients are a factor two to five lower than EO coefficients achieved by parallel-plate poling of bulk material when using polymer-host claddings. For YLD124, an r_{33} of roughly 30 pm/V is reported [70], which leads to $U_\pi = 3$ V for the above mentioned SOH example with an energy consumption of 0.9 pJ/bit. The small in-device EO coefficients r_{33} are due to a lower efficiency in EO dipole re-orientation during poling, as can be seen from Eq. (2.24).

By the beginning of this thesis, work on an SOH phase modulator was published [31]. These modulators were, for the first time, capable of operating at high data rates of up to 42.7 Gbit/s, while modulation in previously reported devices was either only demonstrated in static experiments, or at frequencies below 1 GHz due to the limited bandwidth of the devices. However, energy consumption of the phase modulator [31] was still far from ideal. With an in-device EO coefficient of only 20 pm/V, power-hungry RF amplifiers were required for providing the large drive voltages, resulting in an energy consumption of several pJ per bit. In addition, optical insertion loss of this structure was 40 dB which required several optical amplifiers at the input and output of the device for operation.

These issues are addressed in this thesis and high-speed SOH MZM operating at data rates of up to 84 Gbit/s are demonstrated [34]. The optical insertion loss was significantly reduced to 16 dB by optimizing passive building blocks of the MZM, such as power splitters and combiners as well as strip-to-slot WG mode converters. The residual loss was dominated by non-optimized fiber-chip-couplers and fabrication imperfections. Drive voltage and corre-

Table 2 - Overview of achieved nonlinearities in SOH devices

EO material	In-device r_{33} [pm/V]	Bulk r_{33} [pm/V]
YLD124 (25wt.%)/ APC	30 [70]	118 [69]
YLD124 (25wt.%) / APC	32 [77]	118 [69]
YLD156 (25wt.%) / PMMA	19 [77]	52 [78]
AJSP100 / PMMA	40 [32]	65 [79]
AJLZ53 (25wt.%) / APC	59 [80]	90 [80]
AJLZ53 (15wt.%) / PMMA	19 [81]	60 [81]
M1	20 [31]	75 [82]

sponding energy consumption was significantly reduced by exploring novel EO organic materials with in-device EO coefficients as high as 230 pm/V [71], leading to energy consumptions of a few fJ/bit [72]. This high-speed, low-loss, integrated SOH MZM proves to be a major building block for the development of versatile PICs. Further development led to integrated IQ-modulators [35], capable of transmitting 16QAM signals at a data rate of 112 Gbit/s. Recently, also integrated SOH modulator-based frequency comb sources [73]–[75] and SOH frequency shifters [76] were discussed.

5.2 Passive Components

In the following, the main passive building blocks of the SOH MZM are presented. Multimode interference couplers (MMI), see Section 5.2.1, are used as power splitters and combiners. Strip waveguides, see Fig. 2.3(a) connect grating couplers, MMIs, and phase modulators. Low-loss strip-to-slot mode converters, see Section 5.2.2, are used to couple strip waveguides to slot waveguides.

5.2.1 Multimode Interference (MMI) Couplers

Multimode interference couplers find widespread use for splitting and combining power in integrated photonic circuits on various material platforms. A detailed theoretical description of MMI couplers can be found in [83].

An $n \times m$ MMI coupler has n input and m output singlemode waveguides. Input waveguides are connected to output waveguides by a multimode waveguide section. The transition between singlemode waveguides and the multimode waveguide section is abrupt. The field distribution at the input of the multimode section excites a set of eigenmodes of the multimode waveguide. The modal excitation coefficients can be calculated by an overlap integral, Eq. (B.20), according to Appendix B.3. These modes have different propagation constants. As a result, the cross-sectional field distribution changes during propagation along the multimode waveguide section. The length of the MMI coupler is designed such that the interference pattern of the eigenmodes at the end of the multimode waveguide section comprises m field maxima of equal field strength. After this length an abrupt transition from multimode waveguide to m singlemode waveguides occurs. The m singlemode output waveguides are aligned to the local field maxima of the multimode section. This way the input power of each of the n input waveguides is distributed evenly to the m output waveguides.

The simulated electric field propagation in an optimized 1×2 MMI and in an optimized 2×2 MMI are depicted in Fig. 5.2. The structures were simulated using the transient solver of the commercial simulation tool CST Microwave Studio. The insertion loss of the 1×2 MMI is 0.1 dB and the backreflections are found to be below -20 dB. In the case of the 2×2 MMI the insertion loss is 0.3 dB and the backreflections are found to be below –20 dB.

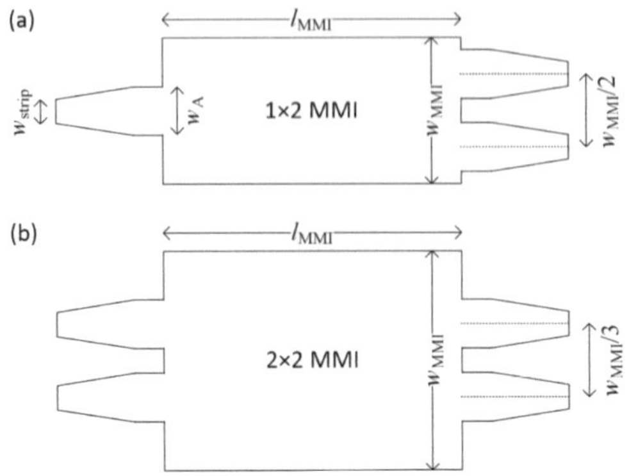

Fig. 5.1 Schematic of multimode interference couplers. (a) 1×2 MMI (b) 2×2 MMI

The MMIs have been fabricated on a 220 nm SOI substrate by IMEC, Belgium, using deep-UV lithography. The measured insertion loss of the 1×2 MMI and of the 2×2 MMI is 0.1 dB and 0.25 dB respectively, in good agreement with the simulations. The losses were extracted from measurements of series of 4, 8, and 12 MMIs. The designed MMIs are used as power dividers and combiners in Mach-Zehnder test structures. The measured extinction ratios are larger than 25 dB as depicted in Fig. 5.3. The demonstrated MMIs are thus well suited for the construction of MZMs.

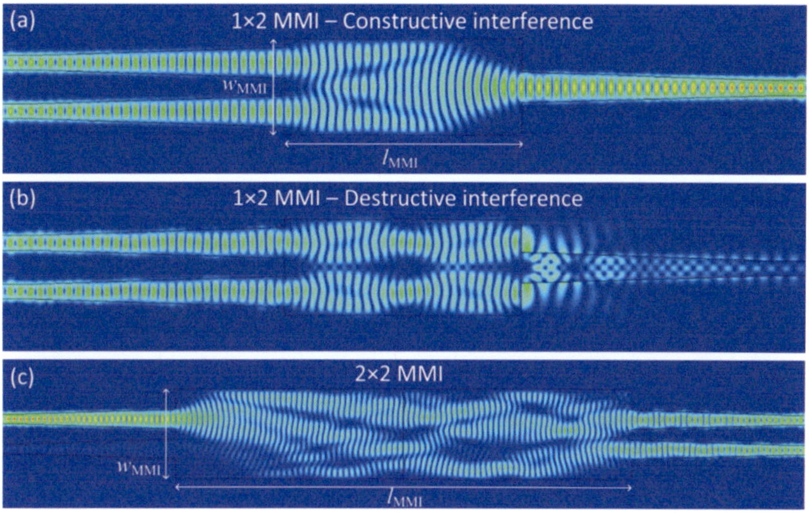

Fig. 5.2 Propagation of the transverse electric field in multimode interference couplers. (a) 1×2 MMI: $w_{MMI} = 3$ µm, $l_{MMI} = 7.5$ µm. Field propagation from right to left depicts the use as a power divider. Field propagation from left to right shows the application as a power combiner (b) Destructive interference in a 1×2 MMI. The input fields have a phase difference of 180°. (c) 2×2 MMI: $w_{MMI} = 4.5$ µm, $l_{MMI} = 23.5$ µm. The access waveguides are uptapered to a width of 1 µm towards the entrance and exit of the MMI. The simulation was carried out at a wavelength of 1550 nm.

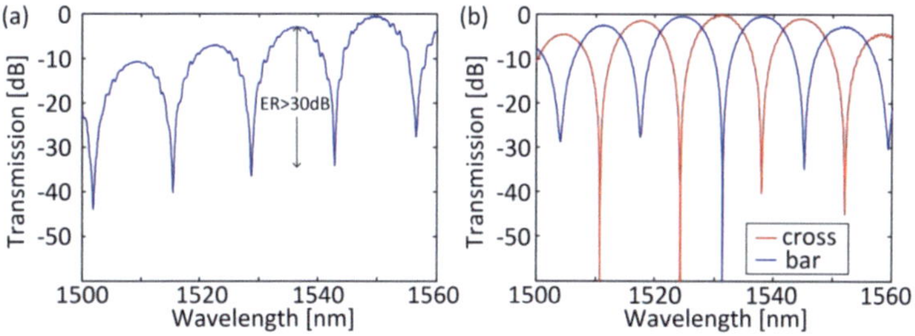

Fig. 5.3 Spectra of unbalanced MZIs based on MMI couplers. (a) MZI using 7.5 µm long 1×2 MMIs. (b) MZI using 23.5z µm long 2×2 MMIs. The measured extinction ratios exceed 25 dB in both cases. The frequency-dependent loss is a result of the grating couplers that were used for fiber-chip coupling and not due to the MMIs.

5.2.2 Low Loss Strip-to-Slot Converter

The following section which reports on mode converters for coupling a strip WG to a slot WG was published in a scientific journal [J4]:

[*Begin of Paper J4*]

Low-Loss Silicon Strip-to-Slot Mode Converters

IEEE Photonics Journal; Volume 5; Issue 1; Article 2200409; January 2013
DOI: 10.1109/JPHOT.2013.2239283

R. Palmer[1,2], L. Alloatti[1], D. Korn[1], W. Heni[1], P. C. Schindler[1], J. Bolten[3], M. Karl[3], M. Waldow[3], T. Wahlbrink[3], W. Freude[1,2], C. Koos[1,2], J. Leuthold[1,2]

[1]*Institute of Photonics and Quantum Electronics (IPQ), Karlsruhe Institute of Technology (KIT), Engesserstr. 5, 76131 Karlsruhe, Germany*

[2]*Institute of Microstructure Technology (IMT), Karlsruhe Institute of Technology (KIT), Hermann-von-Helmholtz-Platz 1, 76344 Eggenstein-Leopoldshafen, Germany*

[3]*AMO GmbH, Otto-Blumenthal-Str. 25, 52074 Aachen, Germany*

We demonstrate compact, highly efficient, broadband strip-to-slot mode converters in silicon with record-low losses of 0.02 (±0.02) dB and negligible reflections between 1480 nm and 1580 nm. The new strip-to-slot transition is logarithmically tapered, which enables a compact design. The new logarithmic tapers are compared to more conventional linearly tapered converters.

Introduction

Strip-to-slot mode converters are key elements for linking well established, strongly guiding strip waveguides to slot waveguides that attracted recent interest for building active silicon devices. Slot waveguides [27], [84] offer new possibilities to functionalize silicon waveguides by hybridization of silicon with highly nonlinear organic materials. While the fundamental material properties of silicon prohibit lasing and Pockels effect based modulation, organic materials can be engineered to complement missing properties of silicon such as optical gain [85],[45] and second-order nonlinearity [29]–[31], [86], [87]. The silicon-organic hybrid approach (SOH) [26], where strongly guiding silicon waveguides are covered with an organic cladding, combines the advantages of both materials. To efficiently exploit the nonlinear properties of the organic cladding, the interaction between guided field and cladding needs to be maximized. Therefore slot waveguides formed by two silicon "rails" with a gap in-between, Fig. 5.4(b), have been found to be superior to more conventional strip waveguides, Fig. 5.4(a). Using slot waveguides covered with a nonlinear organic cladding, high-data rate all-optical signal processing has been demonstrated [43], [44]. Also, energy-efficient liquid crystal phase shifters [88] and high-speed modulation exploiting the ultrafast Pockels effect [29]–[31], [86] take advantage of slightly modified, so-called "striploaded" slot waveguides, Fig. 5.4(c). These *n*-doped strips adjacent to the rails provide an electrical connection to RF transmission lines which control a strong electric field inside the 100 nm wide slot filled with the organic cladding. The strips act as an RF "load" for the slot waveguide capacitance. However, most of the components realized in silicon photonics are based on strip waveguides that have low propagation loss of 2 dB/cm, even for small bend radii [3], while slot waveguide losses are usually in the order of (7…20) dB/cm [89]. Therefore an efficient strip-to-slot waveguide transition is needed to exploit the advantages of each waveguide type. Efficient strip-to-slot mode converters are therefore key components.

Dependent on the application the goal is to couple strip waveguides, Fig. 5.4(a), to slot waveguides, Fig. 5.4(b), or strip waveguides to striploaded slot waveguides, Fig. 5.4(c). Such a converter between the waveguide types needs to have low losses as well as low reflection to avoid multi-path-interference. If a strip waveguide should be coupled to a striploaded slot waveguide, in addition, the two rails of the slot waveguide need to be electrically

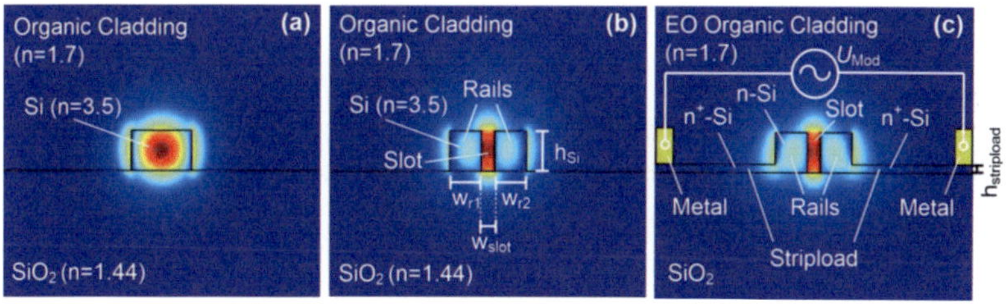

Fig. 5.4 Different types of silicon waveguides. (a) Strip waveguide with dimensions of 220 nm x 450 nm. The light is strongly confined in the silicon strip. (b) Slot waveguide with a rail width of $w_{r1,2} = 240$ nm each and a slot width of $w_{slot} = 100$ nm. The light is strongly confined in the slot. This allows to efficiently exploit the optical properties of an organic cladding material. (c) Strip-loaded slot waveguide of same dimensions as in (b). Metal electrodes are connected to the two rails of the slot waveguide by doped 45 nm high silicon strips (stripload). The slot is filled with an organic electro-optic material. An applied alternating voltage (U_{Mod}) drops essentially across the slot. The resulting strong electric field effectively changes the refractive index, thus enabling modulators with low π-voltage.

insulated as each of the slot wave guide rails is in electrical contact with an electrode having a different potential [31]. Previously proposed strip-to-slot converters were based on linear tapers [29], [90]. A Y-shaped converter has been proposed by Brosi et al. [29] and a loss of 0.13 dB has been reported meanwhile for a 9 µm long converter [90]. The fabrication of such a converter however is challenging for optical lithography, since it requires sub-100 nm features. Recently, a short strip-to-striploaded slot converter which is more suitable for photolithography has been demonstrated based on a spline taper with a loss of 0.8 dB for a 5 µm long converter [91].

We experimentally demonstrate, to the best of our knowledge, the first strip-to-slot mode converters with less than 0.05 dB insertion loss, 0.02 dB in the best case, for converters with length of 7.5 µm and 16 µm respectively. The average loss of the shortest fabricated converter of 4.5 µm length was only 0.26 dB per converter. The converters are based on a sophisticated design consisting of three taper sections that are independently optimized. Logarithmic tapers are used to reduce the length of the converters. The design avoids sub-100 nm features making the design compatible to 193 nm deep UV lithography. The design combines all the advantages of ease of fabrication, electrical insulation in combination with ultra-low reflections and short length. We believe it is a key component for versatile silicon-organic hybrid devices combining the advantages of high-index contrast waveguides and specially engineered organics.

Mode Converter Design

For converting one modal field into a different one, adiabatic transitions are desired. Logarithmically tapered waveguide transitions have been found to have lower conversion losses and reduced length as compared to linear transitions, as will be shown later.

Fig. 5.5 depicts two schemes of mode converters. Fig. 5.5(a) shows a converter for coupling a strip waveguide, Fig. 5.4(a), to a slot waveguide, Fig. 5.4(b). The width of the two rails of the slot waveguide is logarithmically tapered in separate sections. Fig. 5.5(b) shows a converter between strip waveguide, Fig. 5.4(a), and strip-loaded slot waveguide, Fig. 5.4(c), as needed for SOH-modulators [26], [31], [87]. The two rails of the slot waveguide are connected to 45 nm high silicon strips (stripload) in this case. A third taper section is added prior to section I to asymmetrically add a stripload to rail 1. An SOH-modulator requires that the two rails of the slot waveguide are electrically insulated, since they are connected to electrodes of different potential. The converter fulfils this requirement and combines it with low insertion loss and low reflections.

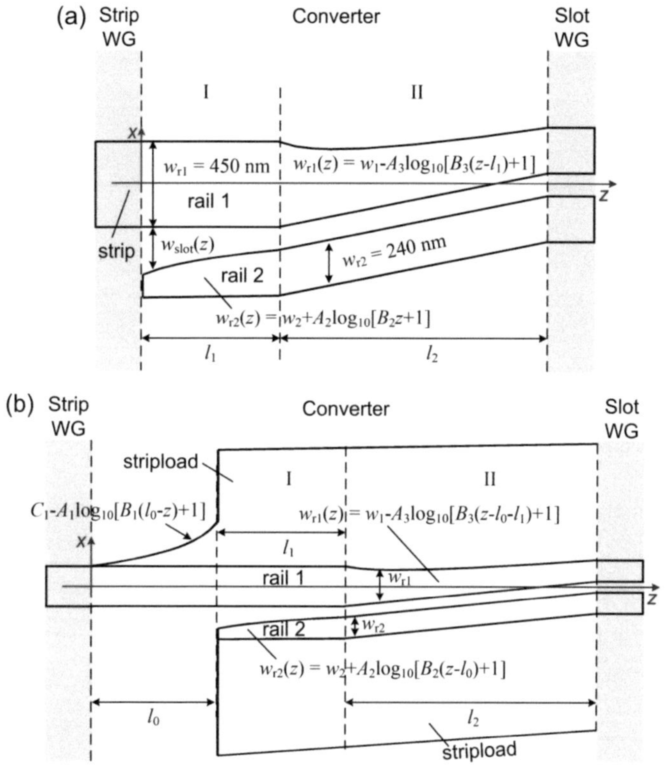

Fig. 5.5 Schematic of the converter based on logarithmic tapers. (a) Strip-to-slot converter. The converter is split in two sections. In section I the width of rail 2 is logarithmically uptapered from 120 nm to 240 nm. Conversely, the width of the slot is simultaneously downtapered from 240 nm to 120 nm. In section II the slot width w_{slot} and the width w_{r2} of rail 2 remain constant, while the width w_{r1} of rail 1 is tapered down from 450 nm to 240 nm. (b) Schematic of a converter between a strip waveguide and a striploaded slot waveguide. Here, to the left of section I, the silicon stripload is uptapered.
Parameters: $A_1 = 500$ nm, $A_2 = 200$ nm, $A_3 = 200$ nm, $B_1 = 2.25\ \mu m^{-1}$, $B_2 = 0.87\ \mu m^{-1}$,

Simulation

In this section the transmission properties of the converter are studied. The choice of the logarithmic shape of the converters is justified, and the logarithmically tapered strip-to-slot converters are compared to linearly tapered converters. In addition, the influence of slot waveguide parameters is investigated.

The Proper Choice of the Taper Shape

Decades before scientists studied the first adiabatic mode transitions of planar optical waveguides the problem of mode conversion was studied intensively by the electrical engineering community in the context of antennas or striplines. The intention was usually to transform one wave impedance of a metallic waveguide into another impedance while simultaneously keeping reflections low. The reflections at the input of such an inhomogeneous metallic waveguide were calculated by approximating the taper with a sequence of short sections of constant impedance and by summing over all partial reflections [93]. To achieve a short transition with low reflections, a variety of taper shapes like parabolic, exponential, hyperbolic cosine and triangular have been proposed.

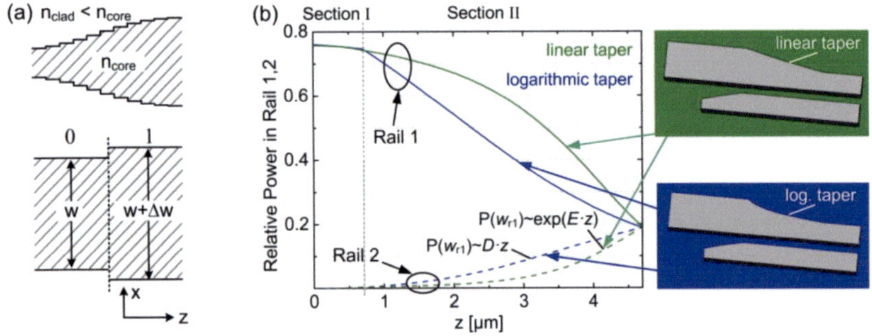

Fig. 5.6 (a) Approximation of a taper by a series of small steps. The width of the structure is only slightly changing from segment to segment. In order to have a nearly adiabatic transition the coupling coefficient between the neighboring segments needs to be close to unity. Modified after [92]. (b) Power guided in rail 1 and rail 2 normalized to the total cross-sectional power. Comparison between linear and logarithmic taper. The figure demonstrates that a logarithmic taper results in a near ideal compact close-to-adiabatic transmission. In the linear case the power guided in rail 2 increases exponentially. Applying the step transition model we therefore expect a steady decrease of the coupling coefficients. For the logarithmic taper the power in rail 2 increases nearly linearly. Therefore we expect an almost constant coupling coefficient between the individual taper segments in the logarithmic case which corresponds to the shortest realization of an adiabatic transition.

The choice of the proper impedance taper function depends on the application in terms of length, allowed reflection and bandwidth, e.g., the exponential taper is among the shortest if a moderate reflection factor is not crucial, while a slightly longer so-called "triangular" [92, Eq.(5.71)] taper results in much lower reflections. For long transitions linear and nonlinear taper functions perform equally well.

Though the same taper functions are often used for dielectric waveguide tapers, the optimization criterion is slightly more restricted. As before, the guided power should be conserved, thus reflection but also radiation of power should be avoided. An approach to solve this problem was introduced by A. Milton and W. Burns [92]. They derived the optimum taper geometry for a transition from a 2 μm wide planar waveguide to a 50 μm wide channel waveguide. The ansatz was to use the step transition model of Marcuse [94], see Fig. 5.6(a). The locally transmitted power from hybrid mode i to mode j at each step of the taper can be calculated by an overlap integral between the modal electric and magnetic field vectors $\vec{\tilde{E}}_{(i,j)(0,1)}$, $\vec{\tilde{H}}_{(i,j)(0,1)}$ to the left (subscript 0) and to the right (subscript 1) of the transition in analogy to the working principle of mode expansion solvers [95]. The power coupling coefficient between local mode i to the left of the transition and local mode j to the right of the transition is given by (B.20) and reference [92]

$$\kappa_{ij}^{(P)} = \frac{\int_{-\infty}^{\infty}\int_{-\infty}^{\infty}\left[\left(\vec{\tilde{E}}_{i0}\times\vec{\tilde{H}}_{j1}^{*}\right)_z+\left(\vec{\tilde{E}}_{j1}^{*}\times\vec{\tilde{H}}_{i0}\right)_z\right]\mathrm{d}x\,\mathrm{d}y}{2\left[\int_{-\infty}^{\infty}\int_{-\infty}^{\infty}\left[\left(\vec{\tilde{E}}_{i0}\times\vec{\tilde{H}}_{i0}^{*}\right)_z\right]\mathrm{d}x\,\mathrm{d}y\int_{-\infty}^{\infty}\int_{-\infty}^{\infty}\left[\left(\vec{\tilde{E}}_{j1}\times\vec{\tilde{H}}_{j1}^{*}\right)_z\right]\mathrm{d}x\,\mathrm{d}y\right]^{1/2}}. \quad (5.1)$$

For a close to adiabatic transition all modes to the left should couple to the equivalent modes to the right with a coupling coefficient close to one while the cross couplings should be as small as possible. In mathematical terms this means $\kappa_{ij}^{(P)} \approx \delta_{ij}$, where δ_{ij} is the Kronecker symbol. Under the assumption that the effective width of the waveguide is equivalent to its physical width, and restricting the design to small taper angles, the authors [92] find that the nearly ideal taper function is a parabola. However, for narrow, strongly confining high index contrast waveguides this assumption does not hold anymore since the evanescent field of the guided mode is in the order of the physical width of the waveguide. Nevertheless for high index contrast structures many other authors found exponential and parabolic tapers to work well [96]. For the strip-to-slot converter under consideration, neither the parabolic taper nor the exponential taper can directly be used because of the following reasons: First, the effective width here is much different from the physical dimensions and second, we consider a transition from a one-rail structure to a coupled two-rail waveguide. However, the criterion formulated by Milton and Burns [92] still holds: The most compact adiabatic transition between two modal field distributions is achieved when the coupling coefficient $\kappa_{ij}^{(P)}$ remains constant and close to unity along the taper length. Therefore, the requirement is that the difference in the modal field distributions of the two sides of each step should be small. In order to get the shortest possible solution, the differential change $\mathrm{d}F(x,y,z)/\mathrm{d}z$ of the modal field distribution $F(x,y,z)$ should be constant such that the coupling coefficient remains constant. Fig. 5.6(b) shows the power guided in rail 1 and rail 2 normalized to the total cross-sectional power while tapering down rail 1 for the case of a linear and a logarithmic taper.

The linear taper is an example for an unoptimized taper. It can be seen from Fig. 5.6(b) that the guided power couples exponentially to rail 2 if the width of rail 1 is linearly downtapered. Therefore the coupling coefficient monotonically decreases along the taper. However, our goal is to achieve a constant differential change in the modal field distribution. This corresponds, to a first approximation, to a linear power exchange between the two rails as rail 1 is downtapered along z. Since the power transfer is found to be nearly exponential along a linear taper, a linear power transfer can be achieved by using the inverse function, namely a logarithmic taper, as depicted in Fig. 5.6(b). It can be seen that for the logarithmic taper the power coupled to rail 2 increases nearly linearly. Therefore we expect that the logarithmically tapered converter is close to ideal.

Linearly Tapered Converter vs. Logarithmically Tapered Converter

Strip-to-slot mode converters based on linear and logarithmic tapers are investigated and compared. It will be shown that a logarithmic taper can outperform a linear taper for short transitions. The minimum length of the converter is mainly determined by taper section II in Fig. 5.5.

The conversion between a strip and a strip-loaded slot waveguide is depicted in Fig. 5.7(a). The figure shows the distribution of the quasi-TE polarized electric field. The performance of the converter has been simulated by a finite integration method using the commercial simulation tool CST Microwave Studio. Our simulations predict losses in the order of 0.01 dB at 1550 nm.

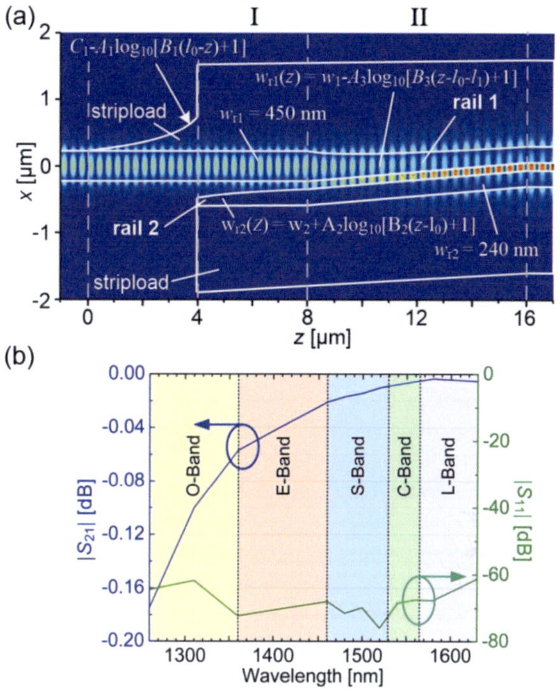

Fig. 5.7 Simulation of a logarithmically tapered converter. (a) Electric field distribution in the logarithmically tapered strip-to-striploaded slot mode converter. To the left of section I a one-sided logarithmic uptaper of the stripload is added. In section I the slot width w_{slot} is logarithmically tapered down from 240 nm to 120 nm (rail 2 tapered up from 120 nm to 240 nm). Section II comprises the logarithmic taper of rail 1 to the final symmetric strip loaded slot waveguide. The slot width and the width of rail 2 remain constant in this section. The simulated loss at 1550 nm is found to be 0.01 dB, while the reflection factor $|S_{11}|$ is below –60 dB. (b) Simulated *S*-parameters of the converter as a function of frequency. Transmission and reflection factor $|S_{21}|$ and $|S_{11}|$. The simulated losses are predicted to be below 0.2 dB (reflection factor below –60 dB) over a wide spectrum comprising the O, E, S, C and L-band.
Parameters: $A_1 = 500$ nm, $A_2 = 200$ nm, $A_3 = 200$ nm, $B_1 = 2.25$ µm^{-1}, $B_2 = 0.87$ µm^{-1}, $B_3 = 1.28$ µm^{-1}, $C_1 = 525$ nm, $w_1 = 450$ nm, $w_2 = 120$ nm, $l_0 = 4$ µm, $l_1 = 4$ µm, $l_2 = 8$ µm

In a wavelength range from 1260 nm to 1675 nm (comprising the O, E, S, C and L-Band) the simulation shows losses below 0.2 dB, and the reflection factor has been found to be below –60 dB.

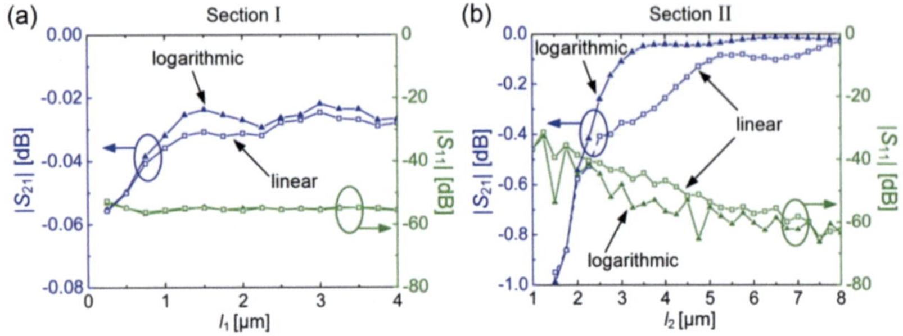

Fig. 5.8 Comparison between linear and logarithmic taper (a) Section I. The length of the taper is varied. In this simulation of section I we chose a section II with 4 µm long logarithmic taper. The transmission is high even for tapers as short as 700 nm. Using a logarithmic taper in section I gives some small advantages. The reflection is nearly constant down to shortest lengths. (b) Section II. The length of the taper is varied. For section I a 700 nm long logarithmic taper has been chosen. For a taper length in section II that is between 2 µm and 8 µm a logarithmic taper results in up to 0.3 dB lower losses as compared to a linear taper of same length. Above a length of 8 µm the linear and logarithmic taper perform alike. The reflection coefficient becomes smaller as the length of the taper is increased. The reflection is slightly lower for a logarithmic taper.

The losses in the two sections of the converter are studied separately in Fig. 5.8(a) and (b). First, the length of section I in Fig. 5.5 is varied while keeping all parameters of section II constant, Fig. 5.8(a). The transmission ($|S_{21}|$) and reflection factor ($|S_{11}|$) are observed. It can be seen that the length of section I has only little impact on the performance of the converter. Loss and reflection are small even for a taper length smaller than 1 µm. This section mainly helps keeping reflections low. The benefit of tapering logarithmically instead of linearly is minor for this section. In contrast, the length of section II has a stronger impact on the transmission properties of the converter, see Fig. 5.8(b). A logarithmic taper outperforms the linear transition by up to 0.3 dB for short taper lengths from 2 µm up to 8 µm. For longer tapers linear and logarithmic tapers perform alike as expected.

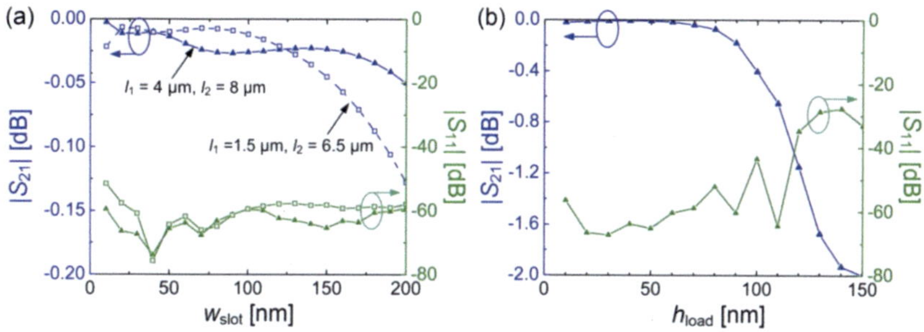

Fig. 5.9 Variation of waveguide parameters. (a) Transmission and reflection of a strip-to-slot converter for different slot sizes. Two converter lengths are investigated. The longer converter outperforms the shorter converter for slots wider than 130 nm. (b) The impact of the stripload height on the performance of the converter is investigated. The converter loss is below 0.2 dB up to a stripload height of 90 nm. This makes the converter robust against fabrication tolerances.

Impact of Slot Waveguide Parameters on Converter Performance

The impact of the slot waveguide parameters on the converter performance is studied. First, the slot width of the logarithmic strip-to-slot converter is varied while keeping the width of the rails constant, Fig. 5.9(a). A 12 μm ($l_1 + l_2 = 4$ μm + 8 μm) and an 8 μm ($l_1 + l_2 = 1.5$ μm + 6.5 μm) long strip-to-slot converter are investigated. Since coupling between the two rails becomes weaker for bigger slots, a better performance is achieved by the longer converter for a slot width bigger than 130 nm. Fig. 5.9(b) shows the influence of the height of the stripload on the transmission properties of the strip-to-striploaded slot mode converter. Its total length is 16 μm ($l_0 + l_1 + l_2 = 4$ μm + 4 μm + 8 μm). The converter loss is below 0.2 dB up to a stripload height of 90 nm.

In conclusion the converter is only weakly affected by a change of the slot width or stripload height. This makes the converter robust against fabrication tolerances.

Fabrication and Characterization

Logarithmically tapered converters between strip and strip-loaded slot waveguides were fabricated on an SOI wafer with a 220 nm silicon device layer and 3 μm thick buried oxide. Converter prototypes were fabricated in the clean-room facilities of AMO GmbH employing AMO's integrated silicon nanophotonic platform. Definition of the photonic devices has been carried out by electron beam lithography (EBL) using hydrogen silsesquioxane (HSQ) as a negative tone resist material. Using HSQ and advanced proximity correction procedure allowed for a precise control of lateral dimensions of the fabricated converters. Pattern transfer into the silicon has been carried out by using a reactive ion etching process based on HBr chemistry. Special care was taken to achieve a high degree of anisotropy and a low sidewall roughness. In total two consecutive process modules, consisting of EBL and HBr-based etching, have been employed for realizing the two different etch depths of the slot waveguide with 45 nm high silicon stripload. Although EBL based fabrication for fast prototyping has been used the design can in principle be realized by using 193 nm deep UV lithography as it does not comprise sub-100 nm features. An SEM image of a fabricated mode converter is depicted in Fig. 5.10(a). The devices were covered with an 800 nm thick layer of PMMA (950k). We characterized several mode converter pairs in series, Fig. 5.10(b,c). The number of subsequent converter pairs was varied. This allows extracting the loss per converter similar to a cut-back measurement. Light was coupled to the chip using grating couplers, identical to the ones reported in [48]. Access waveguides had a constant length. In that way the extracted loss per converter was not affected by the access waveguides. The measured loss therefore comprises the transition loss and the propagation loss of the converter.

A first set of three nominally identical samples has been fabricated. A tuneable laser source and an optical spectrum analyzer were used to measure the transmission spectrum of the devices in a wavelength range between 1480 nm and 1580 nm. As can be seen in Fig. 5.11(a), the spectra of the series connected mode converters do not exhibit super-imposed oscillations, even in the case of 60 converters in series, underlining that the reflection coefficient of

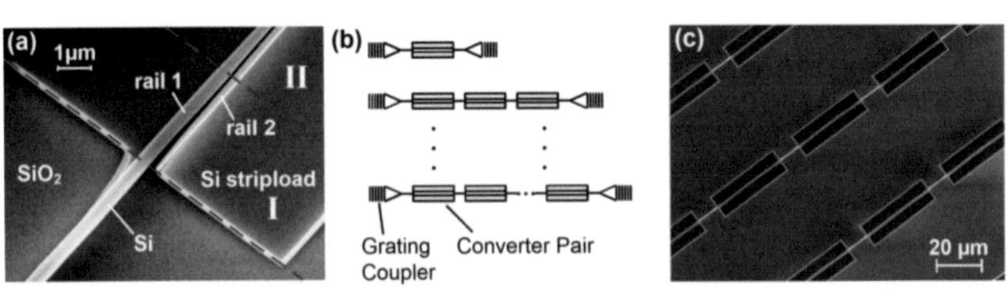

Fig. 5.10 Fabricated test structures. (a) SEM image of a fabricated strip-to-striploaded slot mode converter. (b) Schematic of the chip layout for "cut-back" measurements. Pairs of converters (from strip-to-slot and back) are arranged in series. The number of converter pairs is varied. Light is coupled to the converters by using grating couplers, as reported in [48]. (c) SEM image of the "cut-back" structures.

the converter is indeed very low. The frequency dependency of the spectra stems from the grating couplers and is not a property of the converters. The loss per mode converter, depicted in Fig. 5.11(c), was extracted by a linear regression of total loss versus number of converters at each wavelength separately, Fig. 5.11(b). The error bars represent the 68% (±σ) confidence intervals. The measured loss agrees well with the simulations.

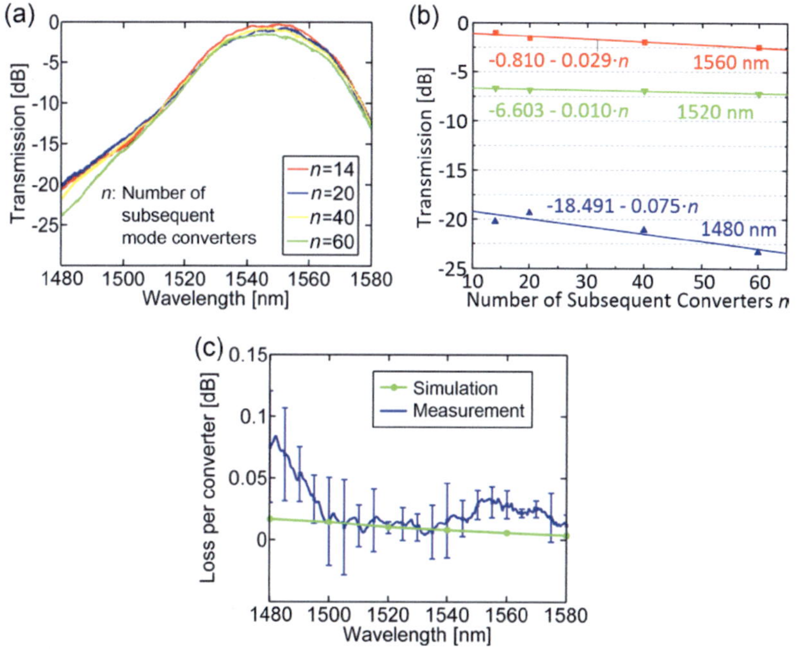

Fig. 5.11 Characterization of the mode converters. (a) Normalized transmission spectra of series of 14, 20, 40 and 60 equally spaced mode converters (pairs of 7, 10, 20 and 30 converters). The spectra do not show any ripples which indicates that there are few if any reflections. The frequency dependency of the spectra is a property of the grating couplers alone [48]. (b) Linear regression of transmission in dB vs. number of converters for three exemplary wavelengths. From the slope of the regression a loss per converter at this wavelength can be inferred. (c) Loss per converter as a function of wavelength. The loss is below 0.1 dB in the entire measured spectral range. The error bars indicate the 68% (±σ) confidence interval. The measurement accuracy was mainly limited by fabrication non-uniformities of the grating couplers. Dimensions: $l_0 = 4$ µm, $l_1 = 4$ µm, $l_2 = 8$ µm

Table 3 - Measured loss of the fabricated mode converters averaged in the wavelength range between 1480 nm and 1580 nm. The error bars indicate the 68% (±σ) confidence intervals. Dimensions: $l_0 = 4$ µm, $l_1 = 4$ µm, $l_2 = 8$ µm

Sample	1	2	3
Average Insertion Loss per Converter [dB]	0.05±0.06	0.04±0.02	0.02±0.02

Table 4 - Measured loss of mode converters of different length averaged in the wavelength range between 1480 nm and 1580 nm. The error bars indicate the 68% (±σ) confidence intervals.

Total length [µm]	**4.5**	**7.5**	**16**
Length of stripload taper [µm]	1.5	1.5	4
Length of section I [µm]	0.5	1.5	4
Length of section II [µm]	2.5	4.5	8
Average Insertion Loss per Converter [dB] Sample 4	**0.18±0.06**	**0.05±0.02**	**0.03±0.04**
Average Insertion Loss per Converter [dB] Sample 5	**0.34±0.06**	**0.05±0.04**	**0.03±0.03**
Simulated Loss per Converter [dB]	**0.21**	**0.05**	**0.01**

Table 3 summarizes the measurement results of three identical chips, averaged over the measured spectral range (1480 nm to 1580 nm). The measured losses were reproducible and well below 0.1 dB, in the best case 0.02 dB per converter. The measurement accuracy was mainly limited by fabrication non-uniformities of the grating couplers. This is, to the best of our knowledge, the lowest reported loss for a mode converter between a strip waveguide and a striploaded slot waveguide.

A fourth and a fifth sample (Sample 4 and 5) were fabricated comprising converters of different lengths. Table 4 shows the measured losses of the converters as well as the expected simulated loss. The shortest converter was 4.5 µm in length and had a loss of 0.18 dB, in the best case, 0.63 dB less than the best previously reported strip-to-striploaded slot converters [91] and slightly shorter. Converters with 7.5 µm and 16 µm length had similar insertion loss of only 0.05 dB and 0.03 dB, respectively, which agrees well with the simulated data of Fig. 5.8(b). The measured converter losses are in agreement with the simulation results.

Conclusion

We demonstrated strip-to-striploaded slot converters with the so far lowest reported losses of (0.02 ± 0.02) dB in the C-band and beyond. The measured spectra prove that the optical reflection coefficient is negligible. In addition, the proposed mode converter keeps the rails of the slot waveguide electrically insulated, making it particularly suited for active silicon-organic hybrid devices [29], [30].

[End of Paper J4]

5.3 Organic Electro-Optic Materials for SOH Integration

As shown in Eq. (2.38), a high EO coefficient r_{33} of the organic cladding material is of high importance for efficient modulation at low driving voltages. According to Eq. (2.24) a high EO coefficient can be achieved by using chromophores with high first-order hyperpolarizability β, by utilizing a high chromophore density N, and by achieving a high acentric order of the chromophore ensemble. However, the chromophore density and average acentric orientation are coupled quantities that cannot be maximized independently [97]: Increasing chromophore density leads to strong electrostatic interactions between the dipolar chromophores, thereby counteracting the desired acentric orientation of the ensemble. For high densities, these intermolecular interactions lead to partial crystallization of the material, to vanishing EO activity, and to increased scattering loss.

The conventional approach to mitigate these interactions is based on using a small chromophore load of typically less than 25 wt.% in an inert polymer host matrix. This allows for the fabrication of amorphous EO films with minimized interaction between chromophore dipoles, but comes at the cost of relatively small chromophore concentrations [30]–[35]. Recent approaches [68] try to increase instead both density and acentric orientation by shape engineering of the chromophore molecules. This way, electro-static inter-chromophore interaction is reduced and the use of neat EO materials is enabled [97].

In this thesis four organic EO materials have been investigated for the hybridization in a silicon-organic hybrid device. Two of these materials are chromophore-polymer guest-host systems. The other two materials consist only of chromophores. Here, an amorphous layer is formed without the use of a polymer-host matrix.

5.3.1 Guest-Host Systems

The most frequently employed class of second-order nonlinear organic claddings for SOH integration are EO molecules, so-called chromophores, which act as dopants in polymeric host-matrices. Prominent examples are YLD124 in PMMA or in APC [70], M3 (commercial product of GigOptix with unknown chemical composition) [82] and AJLZ53 in PMMA or in APC [80], [81].

In this thesis two guest-host chromophore systems are investigated. The first material is a load of 25 wt.% YLD124 in a PMMA matrix, the second material is the commercial EO material M3.

YLD124 / PMMA

The chromophore YLD124 is developed by the research group of Larry Dalton at the University of Washington, Seattle, USA. The chemical structure of YLD124 is depicted in Fig. 5.12(a). YLD124 is a one-dimensional chromophore that consists of a donor group, a π-conjugated bridge and an acceptor group. This basic chromophore structure was discussed in detail in Section 2.1.4. The chromophore is doped at a load of 25 wt.% into a PMMA host matrix to supress strong inter-dipole electro-static interactions. The glass-transition temperature of the guest-host composition is approximately 105 °C. An EO coefficient as high as 118 pm/V has been demonstrated in a bulk reference sample [69]. The highest in-device EO coefficient that was achieved in this work was 30 pm/V, see Section 5.3.3.

M3

M3 is a commercially available EO material supplied by the company GigOptix Bothell, Washington, USA. The chemical composition of this material is proprietary and hence unknown. GigOptix is selling Telcordia certified polymer MZM and IQ-modulators that are based on the material M3. The material shows good long-term stability. The EO coefficient of the material changes by less than 10% at a temperature of 85 °C during a period of 25 years [82]. The glass-transition temperature of M3 is 168 °C. EO coefficients as high as 94 pm/V have been demonstrated using bulk material. In-device values of the commercial polymer modulators range from 60 pm/V to 70 pm/V [82]. The highest EO coefficient that was achieved in an SOH device in this thesis was 40 pm/V. Due to its good thermal stability this material was used for most of the data transmission experiments that are presented in this thesis, see e. g., Section 5.6 and Section 5.7.1.

5.3.2 Monomeric Materials

DLD164

The chromophore DLD164 is developed by the research group of Larry Dalton at the University of Washington, Seattle, USA. The chemical structure of DLD164 is shown in Fig. 5.12(b) and its synthesis is reported in [69]. It consists of a chromophore core (marked in red, donor, π-bridge, and acceptor) which is, despite of a solubilizing attachment to the terminal donor group, identical to the chromophore YLD124, see Fig. 5.12(a). Attached to the EO

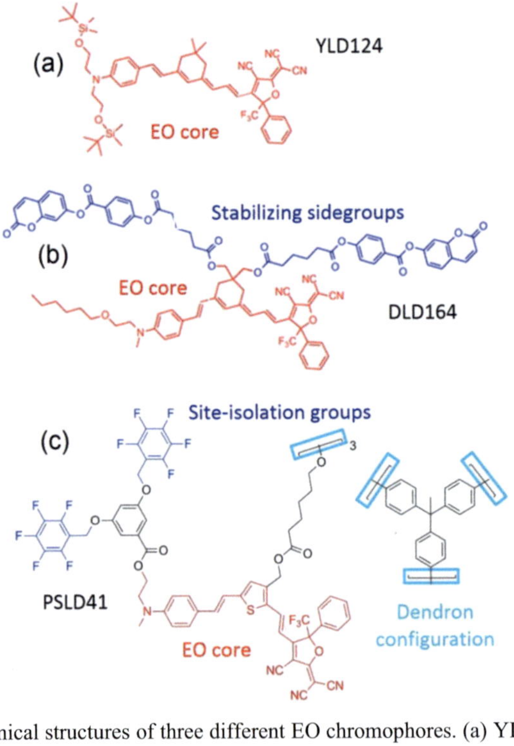

Fig. 5.12 Chemical structures of three different EO chromophores. (a) YLD124 – Due to its strong dipole moment this chromophore is usually embedded in a polymer matrix to avoid micro-crystallization of the material. (b) DLD164 – The molecule comprises an EO active core (red) and pendant stabilizing sidegroups (blue) that suppress dipole-dipole interaction. (c) PSLD41 is a dendritic molecule that combines three EO substructures, marked in light blue. Each EO substructure consists of an EO active region (red) and fluorinated benzene rings (blue) that act as site-isolation groups to suppress dipole-dipole interaction. Modified after [J13]. Copyright © 2014 IEEE

core are two coumarin-containing pendant sidegroups that reduce the dipole-dipole interaction of neighboring molecules. Therefore chromophores can be deployed as a monolithic material, i.e., the material forms an amorphous film without the addition of an isolating matrix. In addition, the coumarin-coumarin interaction between two neighboring chromophores reduces the number of rotational degrees of freedom of the molecules from three to two. As explained in [28] and in Section 5.7.2, this results in a roughly twofold enhancement of the average acentric order parameter in Eq. (2.24) and thus in an enhancement of the EO coefficient as compared to a chromophore without coumarin-containing pendant sidegroups.

The glass-transition temperature of DLD164 is 66 °C. The low glass transition temperature leads to thermally activated reorientation of the dipolar chromophores at room temperature within weeks. The in-device EO coefficient of DLD164 is 180±20 pm/V, which is even higher than the EO coefficients that were achieved in the corresponding bulk reference samples (137 pm/V [69]). Details on the poling experiments can be found in Section 5.3.3 and Section 5.7.2.

PSLD41

The multi-chromophore dendritic molecule (Dendron) PSLD41 was developed by the research group of Larry Dalton at the University of Washington, Seattle, USA and is depicted in Fig. 5.12(c). The dendron consists of a central connecting motif to which three EO substructures are attached. Like DLD164, PDLS41 is a structurally engineered molecule that is optimized for enhanced poling efficiency. However, the dendritic chromophore PSLD41 uses a different approach. Here, perfluoraryl-containing sidegroups (marked in blue) are used to effectively reduce the dipole-dipole interaction of neighboring molecules. The overall substitution pattern results in a spherical shape of the molecule, which is improving the poling efficiency when compared to oblate and prolate structures [97]. The dendron has a moderate glass-transition temperature of 103 °C and the reported EO coefficients of bulk reference samples are as high as 90 pm/V [97]. It is found that very high EO coefficients can be achieved by using PSLD41 as a host-material for YLD124 [98]. This way, 285 pm/V have been reported for parallel-plate poled bulk material. In this thesis a record in-device EO coefficient of 230 pm/V is achieved based on this approach.

5.3.3 Material Comparison

In the following, the poling results of three different EO material concepts are compared in an SOH MZM. The functionalized SOH MZM allow for small device footprint and support high data rates of up to 40 Gbit/s. This section was published in the proceedings of a scientific conference [C32]:

[Begin of Paper C32]

High-Speed Silicon-Organic Hybrid (SOH) Modulators with 230 pm/V Electro-Optic Coefficient Using Advanced Materials

Optical Fiber Communication Conference 2014, San Francisco (CA), USA

R. Palmer[1], S. Koeber[1], M. Woessner[1], D. L. Elder[2], W. Heni[1], D. Korn[1], H. Yu[3,4], M. Lauermann[1], W. Bogaerts[3], L. R. Dalton[2], W. Freude[1], J. Leuthold[1,5], C. Koos[1]

[1]*Karlsruhe Institute of Technology (KIT), Institutes IPQ and IMT, Engesserstr. 5, 76131 Karlsruhe, Germany*

[2]*University of Washington, Department of Chemistry, Seattle, WA 98195-1700, United States*

[3]*Ghent University – IMEC, Photonics Research Group, Department of Information Technology, Gent, Belgium*

[4]*now with: Zhejiang University, Department of Information Science and Electronic Engineering, Hangzhou 310027, China*

[5]*Electromagnetic Fields Laboratory, Swiss Federal Institute of Technology (ETH), Zurich, Switzerland*

We report on record-high electro-optic coefficients of up to 230 pm/V in silicon slot waveguide modulators. The modulators allow for low drive voltage at 40 Gbit/s at a device length of only 250 μm.

Introduction

Silicon-organic hybrid (SOH) devices combine the advantages of strongly guiding silicon photonic waveguides with the wealth of optical properties accessible by molecular engineering of organic materials. In particular, efficient phase modulation can be achieved in SOH devices by using organic cladding materials with pronounced electro-optic (EO) activity. So far, the most commonly used EO materials for SOH integration were polymers doped by small percentages of EO chromophore molecules [68],[99]. However, while these guest-host materials have shown EO coefficients as high as 118 pm/V in bulk material [69], values measured in SOH devices were so far limited to approximately 60 pm/V [99]. Small in-device EO coefficients remained an unsolved issue, preventing the use of the full potential of SOH integration. These limitations have recently been overcome by using a novel class of so-called monolithic EO materials [97], which do not require a polymer matrix to prevent dipole-dipole interaction of chromophores that would lead to vanishing macroscopic EO activity. Such a monolithic material (DLD164) has been applied to an SOH Mach-Zehnder modulator (MZM), resulting in a strong increase of the in-device EO-coefficient to 180 pm/V, allowing for switching energies as low as 1.6 fJ/bit [72].

In this work we demonstrate that even higher in-device EO coefficients of up to 230 pm/V can be achieved by a mixture of traditional EO chromophores and multi-chromophore dendritic molecules (dendron). The extraordinarily high EO coefficient results in a voltage-length product $U_\pi L$ as small as 0.5 Vmm measured at DC. This enables 250 µm long SOH Pockels effect modulators driven by source voltages as small as 2.1 V_{pp} at data rates of up to 40 Gbit/s, where an ER of 9 dB is achieved. We benchmark the results by comparison with a guest-host EO material and with two different monolithic EO materials.

Silicon-Organic Hybrid Modulator

The MZM used in this work is illustrated in Fig. 5.13(a). It consists of two SOH phase-modulators, Fig. 5.13(b), that are driven in push-pull configuration by a single coplanar transmission line (GSG). The copper electrodes of the transmission line are connected to the phase-modulators by 900 nm high tungsten vias. A schematic of a SOH phase modulator along with the electric field of the fundamental quasi-TE mode is depicted in Fig. 5.13(b). The phase modulator consists of a slot waveguide that is electrically connected to an RF transmission line by 60 nm thick doped (As, 7×10^{17}cm^{-3}) silicon strips. The voltage applied to the transmission line drops across the narrow slot, resulting in a high electric field. At the same time, the optical mode is strongly confined to the slot region, resulting in a high overlap with the modulating field. The waveguide is covered and the slot is filled with an electro-optic organic material that is deposited by spin-coating. Directly after the deposition, the organic material does not feature any macroscopic EO effect due to the random orientation of the chromophores. A poling procedure is necessary to induce a macroscopic EO activity. This is accomplished by heating the sample to the

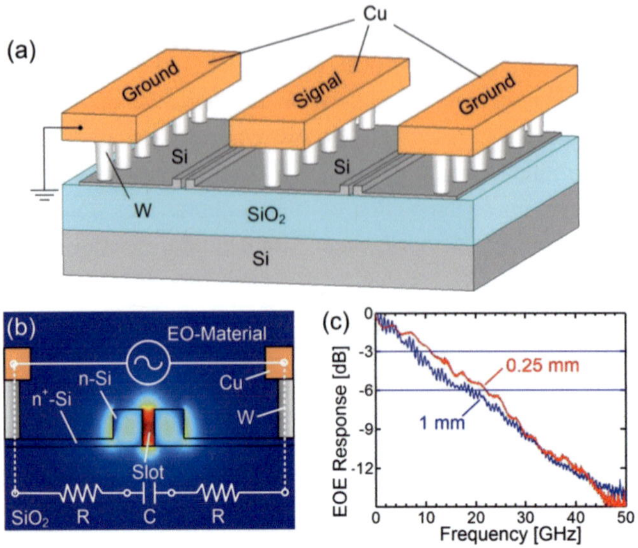

Fig. 5.13 Silicon-organic hybrid MZM. (a) Schematic of the MZM. The device consists of two phase modulators driven in push-pull operation by a single coplanar transmission line (ground-signal-ground). (b) Schematic and simulated optical mode of one arm of the MZM. The two rails of the silicon slot waveguide are electrically connected to metal electrodes by 60 nm high *n*-doped (As, 7×10^{17}cm^{-3}) silicon slabs. The modulation voltage drops across the narrow slot resulting in a high modulation field that has a strong overlap with the optical mode. The silicon slot is filled with an electro-optic organic material. (c) Measured electrical-optical-electrical (EOE) response of a 1 mm long terminated MZM and of a 250 µm long non-terminated MZM, both covered with the organic material mixture YLD124/PSLD41. The 6 dB EOE bandwidths are 18 GHz and 22 GHz, respectively.

glass-transition temperature T_g of the organic material while applying an electric DC field to align the dipolar molecules [33]. While holding the poling voltage, the chip is cooled back to room temperature, thereby conserving the acentric order of the chromophores.

The devices are fabricated by deep-UV lithography on a 220 nm SOI wafer with 2 µm buried oxide. The measured slot width is 160 nm and the rail width is 210 nm. The fiber-to-fiber loss is 16.5 dB, dominated by fiber-to-chip coupling losses of the non-optimized grating couplers. The on-chip loss is approximately 6 dB for maximum transmission of the modulator. Two SOH MZM of different lengths were investigated: Fig. 5.13(c) depicts the measured bandwidths of a 1 mm long modulator terminated with 50 Ω, and of a 250 µm long non-terminated modulator. The 6 dB electrical-optical-electrical (EOE) bandwidths are 18 GHz and 22 GHz, respectively. Applying a gate voltage between the substrate and the transmission line allows to further boost the bandwidth as reported in [31].

Poling Results of the Investigated Electro-Optic Organic Material

The materials investigated in this work are the monolithic chromophore DLD164 [72], the multi-chromophore dendritic molecule PSLD41 [98], the guest-host material YLD124 (25 wt.%) in PMMA [68], and a mixture of YLD124 and PSLD41 (25:75 wt.%). The respective chemical structures are depicted in Fig. 5.14. DLD164, Fig. 5.14(b), and PSLD41, Fig. 5.14(d), are structurally engineered molecules optimized for enhanced poling efficiency. For DLD164, pendant coumarin-containing sidegroups mitigate dipole-dipole interaction and are used to reduce the rotational degrees of freedom of the chromophores from three to two. This improves the degree of chromophore alignment for a given poling field by roughly a factor of two [28]. The dendritic chromophore PSLD41 uses a different approach. Here, perfluoraryl-containing sidegroups are used to effectively reduce the dipole-dipole interaction of neighboring molecules. The dendron chromophore consists of a central connecting motif to which three EO substructures are attached, see Fig. 5.14(d). This results in a spherical shape of the molecule and improved poling efficiency [97]. The investigated materials are applied to nominally identical chips by spin-coating from an 8% solution dissolved in 1,1,2-trichloroethane. After deposition, the materials are poled for a later push-pull operation of the MZM by applying a poling voltage between the ground electrodes of the modulator [33]. After poling, the π-voltage U_π of the modulator is measured at DC so that the EO coefficient can be derived. Fig. 5.14(a) depicts the resulting EO coefficients as a function of the applied poling field for the various materials. The first investigated material is the guest-host system YLD124/PMMA. Similar to previously reported results in other guest-host systems [31], [33], [99] we reach only small r_{33}-values (≤ 29 pm/V) and a low poling efficiency r_{33}/E_{poling} of only 0.23 nm²/V² measured for small poling fields. This is far below the value achieved in the corresponding bulk material [68]. Next, we study the monolithic chromophore DLD164. We find a much higher poling efficiency (1.17 nm^2/V^2) and a large r_{33} of up to 190 pm/V. Finally, we perform poling experiments with a sample coated with pure PSLD41 and with a sample coated with the binary chromophore organic glass YLD124/PSLD41. For PSLD41, we find a remarkable in-device r_{33} of 97 pm/V (poling efficiency 0.31 nm²/V²), while the composition YLD124/PSLD41 was also found to have an r_{33} coefficient that exceeds

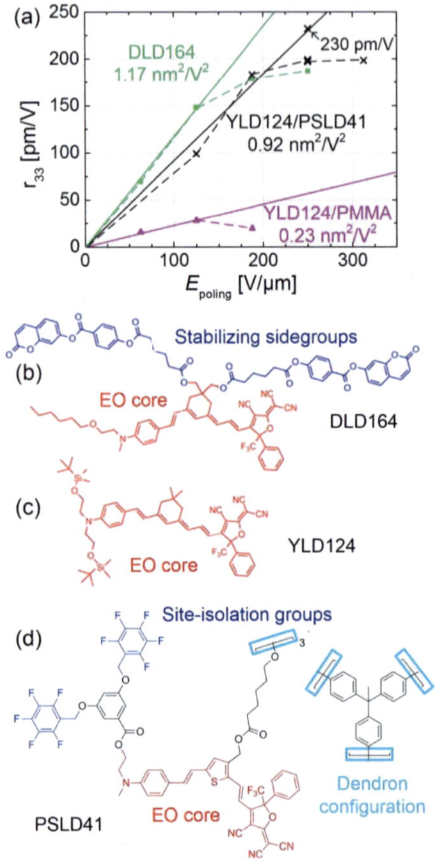

Fig. 5.14 Poling of EO organic materials. (a) Measured poling efficiencies r_{33}/E_{poling} for three different organic cladding materials: The monolithic chromophore DLD164, PMMA doped with 25%wt. YLD124 (guest-host system), and the dendritic chromophore PSLD41 doped by 25%wt. YLD124 (binary chromophore organic glass). The materials DLD164 and PSLD41 consist of structurally engineered chromophores that enable high chromophore densities and a high degree of chromophore orientation during poling. We find extraordinarily high in-device electro-optic coefficients for DLD164 (190 pm/V) and YLD124/PSLD41 (230 pm/V). (b-d) The chemical structures of the individual chromophores are depicted. The EO-cores of the chromophores are drawn in red. Sidegroups that lead to matrix stabilization and enhancement of molecular orientation are marked in blue. PSLD41 is a dendritic configuration that combines three EO substructures, marked in light blue in (d).

Table 5 - Summary of measured EO coefficients. Comparison between in-device values and values achieved in parallel-plate poled bulk reference samples.

Material	n	r_{33} in-device [pm/V]	r_{33} bulk [pm/V]	$n^3 r_{33}$ in-device [pm/V]	$U_\pi L$ [Vmm]	T_g [°C]
YLD124(25wt.%) /PMMA	1.7	29	155 [68]	142	4.10	105
DLD164	1.83	190	137	1164	0.48	66
PSLD41	1.72	97	90 [98]	494	1.23	103
YLD124(25wt.%) /PSLD41	1.73	230	285 [98]	1130	0.52	97

even the sum of the r_{33} coefficients of its constituents [98]. This is the highest reported EO coefficient in an SOH device, and is even higher than the previously reported record value of a fully organic MZM, where 137 pm/V were measured [100]. Table 5 summarizes the results of the poling experiments of the four investigated materials. We see that in contrast to guest-host systems like YLD124 in PMMA, where in-device EO coefficients are usually much smaller than reference values of bulk materials, structurally engineered chromophores like DLD164 or PSLD41 allow for values close to or even higher than the bulk-material reference. This is partially due to the fact that the materials exhibit higher stability with respect to dielectric breakdown in SOH slot waveguides than in bulk references. We attribute this to thin-film effects and low defect densities in the slot region. Note that the in-device $n^3 r_{33}$ figures of merit of the reported organic materials exceed 1100 pm/V and are clearly superior to those of conventional EO materials such as GaAs ($n^3 r_{33} \approx 60$ pm/V) and $LiNbO_3$ ($n^3 r_{33} \approx 400$ pm/V).

Demonstration of 40 Gbit/s OOK

Finally, we perform high-speed modulation experiments based on 250 µm long non-terminated MZM modulators coated with the mixture YLD124/PSLD41. Light from a tunable laser source at 1540 nm is coupled to the modulator, which is biased to its quadrature point. The device is connected to a pattern generator adjusted for a peak-to-peak voltage of 2.1 V_{pp} if terminated with 50 Ω. Reflections of the RF wave at the end of the non-terminated device result in nearly a doubling of the in-device drive voltage to roughly 4.2 V_{pp}. The modulated light is amplified using an EDFA and received using a digital communications analyzer along with a bit error ratio (BER) tester. Fig. 5.15 shows recorded eye diagrams for data rates from 25 Gbit/s to 40 Gbit/s. We

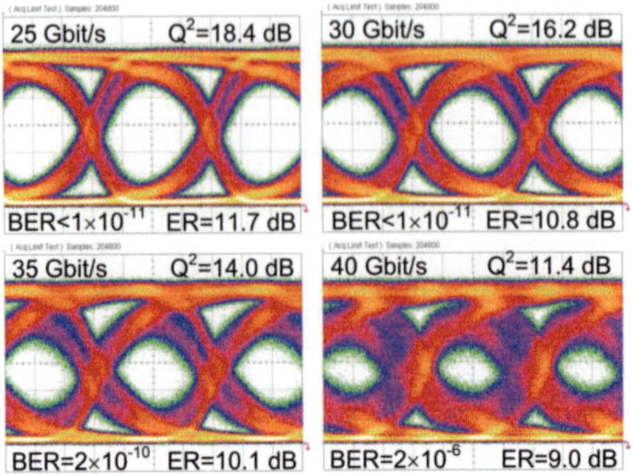

Fig. 5.15 NRZ OOK eye diagrams at 25 Gbit/s, 30 Gbit/s, 35 Gbit/s, and 40 Gbit/s obtained from a 250 µm long MZM without termination. Measured Q^2-factors, extinction ratios (ER) and bit error ratios (BER) are denoted in the figure. Drive voltage across the slot is 4.2 V_{pp}. The ER exceeds 10 dB up to 35 Gbit/s. At 40 Gbit/s we measure a BER of 2×10^{-6}. PRBS length 2^{31}-1. No gate voltage has been used.

observe excellent signal quality and extinction ratios (ER) exceeding 10 dB up to a data rate of 35 Gbit/s. At 40 Gbit/s we measure a low BER of 2×10^{-6} and an ER of 9 dB. The modulator has a measured capacitance of 95 fF resulting in an energy consumption of 420 fJ/bit [72]. It is possible to reduce the energy consumption to a few fJ/bit by increasing the modulator length to 1 mm as reported in [72]. Note that, in contrast to earlier demonstrations of 40 Gbit/s SOH modulators [31],[72], we did not use a gate voltage to improve the silicon conductivity [31]. Still, a small voltage-length product of 1 Vmm is found for operation at 40 Gbit/s, one order of magnitude below typical values reported for reverse-biased *p-n*-modulators [17].

Conclusion

We demonstrate that EO coefficients as high as 230 pm/V can be achieved by hybridizing silicon slot waveguides with structurally engineered electro-optic organic materials. This allows for small voltage-length products of 0.5 Vmm, enabling non-resonant 250 µm long devices operating at 40 Gbit/s.

[End of Paper C32]

5.4 RF-Design

The proposed SOH MZM structure, see Fig. 2.11, is based on striploaded slot waveguides. The striploaded slot WG is both a WG for the optical carrier frequency and for the modulating RF frequency, which has to be considered when designing the RF transmission line. Key parameters for the RF design are the transmission line impedance, the RF attenuation and the group velocity of the RF signal.

5.4.1 Transmission Line Impedance

The impedance of the modulator is targeted to be matched to the standard 50 Ω impedance of an RF source. For a single phase modulator (signal-ground configuration) a 50 Ω line impedance is realized using 25 µm wide signal and ground electrodes and a slot width of 160 nm. In the case of a single-drive push-pull MZM, the RF electrodes connect to the two phase modulators in parallel, as depicted Fig. 2.12. Thus, the impedance of the MZM is only half of the impedance of a single phase modulator, when using electrodes of same width. The RF impedance of the MZM is predominantly determined by the width of the silicon slot and by the width of the signal electrode. The width of the silicon slot needs to be smaller than 200 nm to ensure both waveguiding and a large field interaction factor Γ, see Eq. (2.35). Therefore, the line impedance can mainly be increased by reducing the width of the signal electrode. Due to design restrictions and due to increasing RF loss, the signal electrode cannot be chosen to be arbitrary small. For a narrow electrode width of 10 µm a low line impedance of 33 Ω is predicted, using the commercial simulation

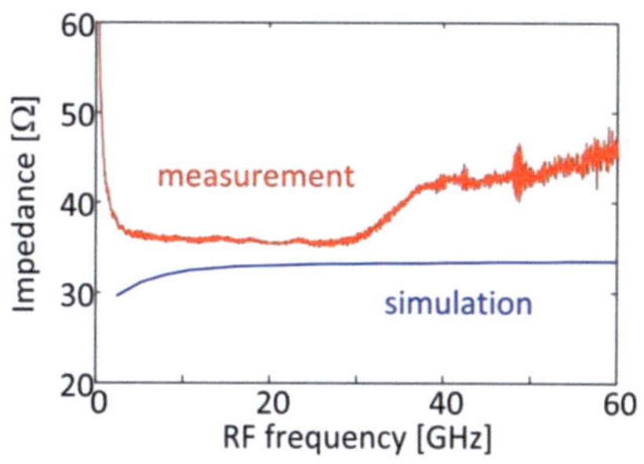

Fig. 5.16 Simulated and measured line impedance of a SOH MZM. The slot width is 160 nm, the width of the signal and ground electrode is 10 µm and 100 µm, respectively, and the separation of the electrodes is 5 µm.

tool CST Microwave Studio. Thus, a voltage reflection factor of 20% is expected, when connecting the modulator to a 50 Ω source. The simulated impedance of an SOH MZM along with the corresponding measured impedance of a fabricated device is depicted in Fig. 5.16. For frequencies up to 25 GHz an impedance of 36 Ω is measured, slightly higher than predicted. For frequencies exceeding 25 GHz a rapid increase of the impedance to roughly 43 Ω can be observed. In this regime the device is RC-limited, see Section 5.5, and the drive voltage U_{drive} no longer predominantly drops across the slot capacitor C_{slot}. Effectively, the slot width increases for frequencies beyond the RC-limit, which results in the observed increase of the line impedance.

In this thesis the low impedance of a single-drive MZM is accepted for achieving high-speed modulation at lowest energy consumption. In principle, impedance matching can be achieved using a dual drive MZM, however, this comes at the price of doubling the dissipated RF power.

5.4.2 Transmission Line Losses

As derived in Appendix D.2 microwave attenuation has its origin in Ohmic losses in the conductor due to the skin effect, and in dielectric loss in the non-ideally insulating substrate. In this section, first, the microwave attenuation of simple "feed" transmission lines on an SOI substrate are investigated, followed by a study of the microwave loss of an SOH MZM.

Feed Line Loss

The investigated feed lines are ground-signal coplanar slotlines with 50 Ω impedance. Aluminum is used as electrode material. SOI wafers are used as substrates. Prior to metal deposition, the silicon device layer is removed by dry etching. Fig. 5.17 compares feed lines which were fabricated on SOI wafers with different substrate resistivity. Furthermore, two different heights of the Al electrodes are investigated.

It can be seen that transmission line loss strongly depends on both electrode thickness and substrate conductivity. Losses below 1 dB/mm were found for frequencies up to 40 GHz when using high resistivity SOI substrates (>1000 Ωcm) and aluminum electrodes of 500 nm height.

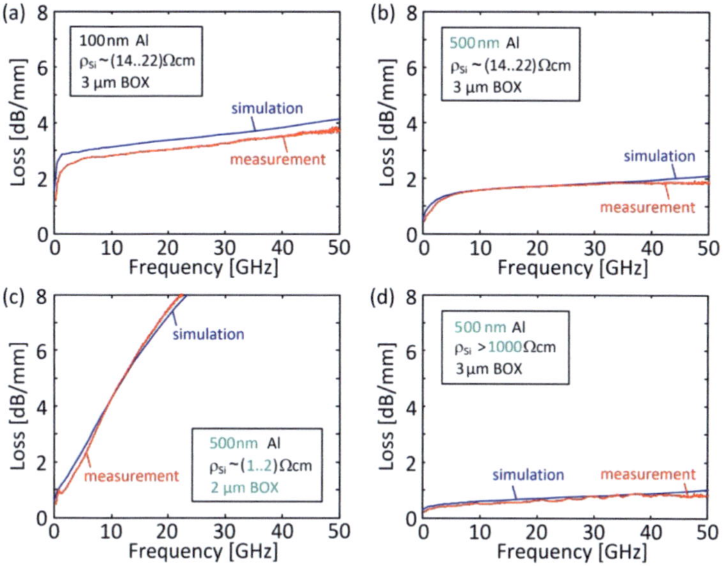

Fig. 5.17 Comparison of RF loss of metallic GS transmission lines on SOI substrates. The top silicon was removed prior to electrode deposition. (a) and (b) depict simulated and measured losses of GS lines of different height deposited on a standard photonic SOI substrate. A reduction of the loss by 1.5 dB/mm is observed when increasing the electrode height from 100 nm to 500 nm. (b)-(d) depict losses of electrodes of 500 nm height on SOI substrates with different resistivity. A strong influence of transmission line loss on the substrate conductivity is observed. Losses below 1 dB/mm up to a frequency of 40 GHz are found when using high resistivity (>1000 Ωcm) SOI substrates and an electrode height of 500 nm. The measured losses are in good agreement with simulations that have been carried out using the commercial simulation software CST Microwave Studio. Samples fabricated by AMO GmbH, Aachen.

Modulator RF Loss

When using SOH modulators based on striploaded slot waveguides the thin doped silicon striploads become part of the transmission line. The conductivity of the striploads is far smaller than the conductivity of metal, and there is little freedom in changing the height of the striploads without sacrificing optical mode confinement. Thus Ohmic losses are expected to be higher for the modulator as compared to purely metallic feed lines. Fig. 5.18(a) shows the simulated and the measured loss of an SOH MZM that was fabricated at IMEC, Belgium, using deep-UV lithography and a CMOS-like metal stack as reported in [35]. The measured loss is significantly larger than predicted for frequencies larger 10 GHz. Also, the measured attenuation appears to scale linearly with frequency for frequencies larger 25 GHz, indicating that dielectric loss dominates the conductor loss, see Appendix D.2. This can be understood as follows: For frequencies larger than 25 GHz the device is operated beyond its RC cut-off frequency, and a significant part of the modulation voltage drops across the resistive stripload. Therefore, part of the stripload acts as a lossy dielectric. Fig. 5.18(b) shows the measured RF loss of an SOH MZM that was fabricated by AMO GmbH, Aachen, Germany using eBeam lithography, and 500 nm high aluminum electrodes as depicted in Fig. 2.15. Loss measurements were carried out while applying various gate voltages U_{Gate}. The application of a positive gate voltage (positive potential of applied to the silicon substrate, see Fig. 2.12) decreases the resistance R_{load} of the silicon stripload by formation of

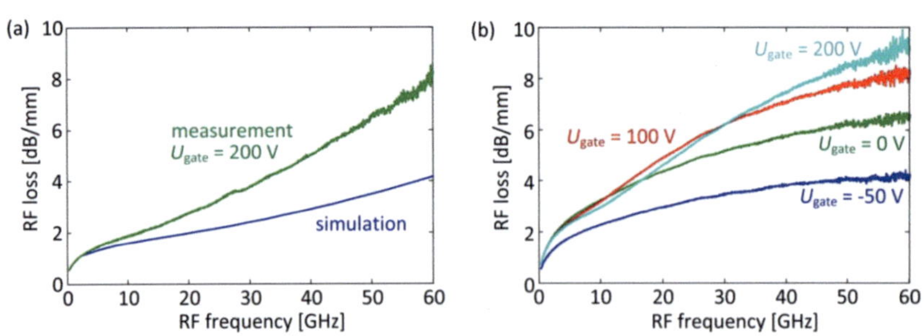

Fig. 5.18 RF loss of an SOH MZM. (a) The device was fabricated by deep-UV lithography using a metal stack as reported in [35]. The targeted stripload conductivity was 25 $(\Omega\mathrm{cm})^{-1}$. (b) The device was fabricated by eBeam-lithography and uses 500 nm thick aluminum electrodes. A gate voltage U_{gate} is used to increase ($U_{\mathrm{gate}} > 0$) or decrease ($U_{\mathrm{gate}} < 0$) the conductivity of the silicon stripload [31]. The measured loss depends strongly on the applied gate voltage.

a highly conductive electron accumulation layer in the silicon striploads [31]. As can be seen in Fig. 5.18(b), the RF loss increases as the (positive) gate voltage increases, i. e. RF loss increases as the stripload conductivity increases. When applying a negative gate voltage, the free charge carriers in the *n*-doped stripload can be depleted. In this case, the stripload behaves like an insulator, and the optical waveguide is decoupled from the electrical transmission line. This corresponds to the case of a metallic feed line with a large separation of the electrodes. Consequently, a reduction of RF loss is expected. This case ($U_{\text{Gate}} = -50\,\Omega$) is depicted by the blue line in Fig. 5.18(b).

5.4.3 Walk-Off

According to Eq. (2.40), a group velocity mismatch between the optical and electrical waveguide fields can cause bandwidth limitations of an EO modulator. The group indices of optical signal n_{g}^{o} and RF signal n_{g}^{m} were simulated using CST Microwave Studio:

$$n_{\text{g}}^{\text{m}} = 2.7, \qquad n_{\text{g}}^{\text{o}} = 3.1. \tag{5.2}$$

As discussed in the following Section 5.5, this difference in group indices has only a minor impact on the frequency response of a 1 mm long device for frequencies below 50 GHz.

5.5 Electro-Optic Bandwidth

In this section, the impact of RF loss, walk-off and RC-characteristics on the EOE bandwidth of the modulator is investigated, according to the analytic model in Section 2.4.3. The predicted frequency response of the MZM is compared to the measured frequency response.

5.5.1 Predicted Bandwidth

In a first step, we use the travelling wave model in Eq. (2.40) to study the impact of velocity mismatch between microwave and optical signal on the modulator response, explicitly neglecting the RC-characteristic of the device. With the simulated group indices, given in Eq. (5.2), the modulation bandwidth of a 1 mm long device decreases by less than 1 GHz as compared to a modulator with perfectly matched group velocities for frequencies below 50 GHz. Walk-off between the two waves can therefore be neglected for such short device lengths.

When neglecting velocity mismatch, the frequency response can be approximated by Eq. (2.49). This model accounts for microwave loss and RC limitations of lumped devices. For stripload resistance and slot capacitance we use measured data of fabricate devices that were coated with the EO material M3 (Section 5.3.1): $R_s = 25\Omega$, $C_{\text{slot}} = \left(-0.33\ln(f_m / \text{Hz}) + 11.11\right) \cdot 41\,\text{fF}$ (C_{slot} = 152 fF at f_m = 5 GHz). The slot capacitance is frequency-dependent due to its dispersive dielectric. Assuming that the RF attenuation is dominated by

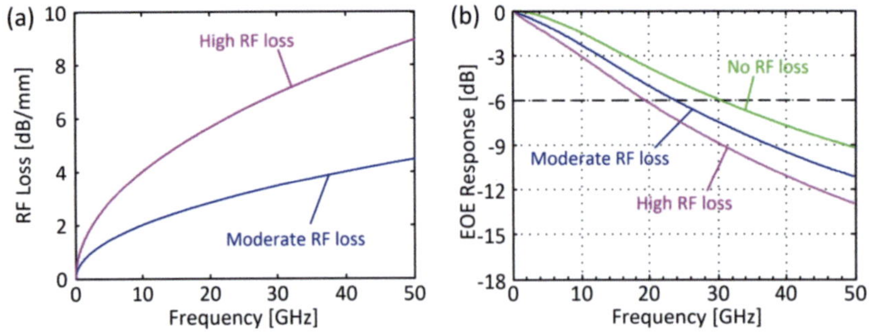

Fig. 5.19 Predicted bandwidth of a single-drive push-pull SOH MZM. (a) Depicted are two exemplary transmission line losses. These two curves have been chosen such that they correspond to the upper and to the lower loss limit of the measured losses in Fig. 5.18. (b) Frequency response of the MZM calculated from the predicted losses shown in (a).

the skin effect, we approximate the frequency-dependency of the microwave power attenuation coefficient $\alpha_{\mathrm{m}} \approx \alpha_{\mathrm{R}'}$ by a square root function as derived in Eq. (D.16).

The influence of microwave attenuation $\alpha_{\mathrm{m}} \approx \alpha_{\mathrm{R}'}$ on the frequency response of the device is investigated. Two exemplary RF losses $\alpha_{\mathrm{R}'}$ are depicted in Fig. 5.19(a). These two loss curves have been chosen such that they correspond to the upper and to the lower loss limit of the measured losses in Fig. 5.18. The loss curves as depicted in Fig. 5.19(a) are substituted in Eq. (2.49). The resulting electrical-optical-electrical (EOE) frequency responses are compared to the case of a modulator with a lossless transmission line in Fig. 5.19(b). The microwave loss reduces the EOE 6 dB-bandwidth of the MZM significantly from 30 GHz to less than 24 GHz.

In conclusion, the bandwidth of a 1 mm long SOH MZM is predominantly limited by the resistive and capacitive lumped elements of source and device, as depicted in Fig. 2.14. In addition, microwave loss of the transmission line reduces the bandwidth by up to 10 GHz. For frequencies below 50 GHz, velocity mismatch between microwave and optical signal can be neglected.

5.5.2 Measured Bandwidth

Next, the model of Section 2.4.3, Eq. (2.49) is verified by comparing the predicted frequency response to the measured frequency response of a fabricated device. The measured EOE response of a 1 mm long MZM is depicted in Fig. 5.20 for different gate voltages. Also here, the EO material M3, see Section 5.3.1, is used as an organic cladding material. The application of a gate voltage (positive potential applied to the substrate silicon, see Fig. 2.12) decreases the resistance R_{load} of the silicon stripload by formation of a highly conductive electron accumulation layer in the silicon striploads. Consequently, the modulator bandwidth increases from 18 GHz to 26 GHz. Details on field-induced electron accumulation can be found in Ref. [31]. For a gate voltage of 150 V we measure a stripload resistance of $R_{\mathrm{load}} = 25\ \Omega$. To accurately predict the device bandwidth we insert the measured frequency-dependent microwave attenuation, as depicted in Fig. 5.18(a), and the measured frequency-dependent slot capacitance $C_{\mathrm{slot}} = \left(-0.33\ln(f_{\mathrm{m}}/\mathrm{Hz}) + 11.11\right)\cdot 41\,\mathrm{fF}$ into Eq. (2.49). A comparison between measured (blue) and predicted (magenta) EOE response

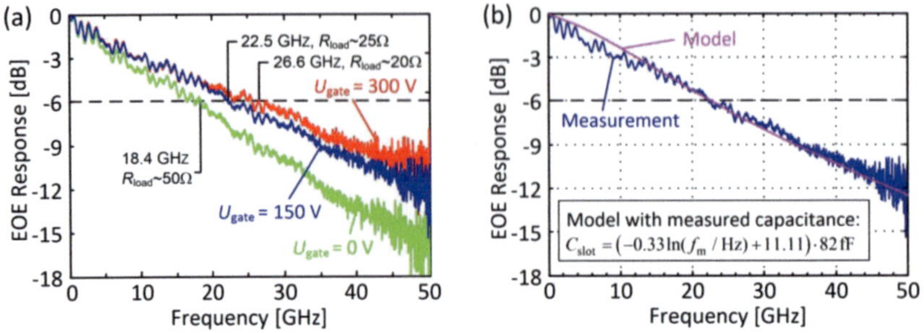

Fig. 5.20 Measured EOE response of a 1 mm long SOH MZM with M3 cladding. (a) The bandwidth increases from 18 GHz to 26 GHz when applying a gate voltage of 300 V. The BOX thickness is 2 μm. (b) Comparison between measurement and analytic model.

is depicted in Fig. 5.20(b). Model and measurement are in good agreement. It can therefore be concluded that the analytic model in Eq. (2.49) can be used to predict the frequency response of the SOH MZM with reasonable accuracy.

5.6 High-Speed Modulation at Data Rates of up to 84 Gbit/s

In this section we demonstrate that the developed SOH MZM supports modulation at high symbol rates and even supports multiple modulation amplitudes. This section was published in a scientific journal [J5] and in part in the proceedings of a scientific conference [C21]:

[Begin of Paper J5]

Silicon-Organic Hybrid MZI Modulator Generating OOK, BPSK and 8-ASK Signals for up to 84 Gbit/s

IEEE Photonics Journal; Volume 5; Issue 2; Article 6600907; April 2013
doi: 10.1109/JPHOT.2013.2258142

R. Palmer,[1] L. Alloatti,[1] D. Korn,[1] P. C. Schindler,[1] R. Schmogrow,[1] W. Heni,[1] S. Koenig,[1] J. Bolten,[2] T. Wahlbrink,[2] M. Waldow,[2] H. Yu,[3] W. Bogaerts,[3] P. Verheyen,[4] G. Lepage,[4] M. Pantouvaki,[4] J. Van Campenhout,[4] P. Absil,[4] R. Dinu,[5] W. Freude,[1] C. Koos,[1] J. Leuthold,[1,6]

[1]Karlsruhe Institute of Technology (KIT), Institutes IPQ and IMT, Engesserstr. 5, 76131 Karlsruhe, Germany

[2]AMO GmbH, Aachen, Germany

[3]Ghent University – IMEC, Photonics Research Group, Department of Information Technology, Gent, Belgium

[4]IMEC vzw. Kapeldreef 75, Heverlee, Belgium

[5]GigOptix Inc., Switzerland and GigOptix Bothell, Washington, USA

[6]Institute of Electromagnetic Fields (IFH), Swiss Federal Institute of Technology (ETH), Zurich, Switzerland

We report on high-speed multilevel signal generation and arbitrary pulse shaping with silicon-organic hybrid (SOH) Mach-Zehnder interferometer (MZI) modulators. Pure phase modulation exploiting the linear electro-optic effect allows the generation of multiple modulations formats at highest speed such as 40 Gbit/s OOK and BPSK and 28 GBd bipolar 4ASK and bipolar 8ASK with data rates up to 84 Gbit/s. Additionally, beside NRZ pulse-shaping, for the first time Nyquist pulse shaping with silicon modulators is demonstrated to enable multiplexing at highest spectral efficiency.

Introduction

Silicon electro-optic modulators are key elements for photonic-electronic integration in telecommunications and optical interconnects. Devices capable of generating advanced modulation formats are of growing importance to increase bit rates while keeping symbol rates moderate and thus compliant with CMOS driver electronics. Commercial devices, relying on the linear electro-optic effect based on material platforms like $LiNbO_3$, so far have shown the best results for the generation of multilevel signals. For lack of a linear electro-optic coefficient, silicon modulators largely use free-carrier dispersion in forward-biased *p-i-n* junctions [18], [101], [102] or reverse-biased *p-n* junctions [17], [20], [23], [103]–[105]. The latter concept enables higher modulation bandwidth [17] and has recently been used to demonstrate 50 Gbit/s quadrature-phase-shift keying (QPSK) with a single IQ modulator [106] and 112 Gbit/s with two IQ modulators and polarization multiplexing [21]. However, free-carrier dispersion leads to an intrinsic coupling of amplitude and phase modulation in all-silicon devices, which may render higher-order quadrature amplitude modulation (M-QAM, $M > 4$) challenging. In addition, so far reported non-resonant reverse biased *p-n* modulators typically have voltage length products $U_{\pi}L$ not better than 10 Vmm and offer moderate extinction ratios in the order of (3 to 8) dB at data rates above 20 Gbit/s [17], [23], while extinction ratios exceeding 10 dB have been reported at lower data rates [107], [108]. A promising candidate to improve these results is silicon-organic hybrid (SOH) integration. SOH devices offer a linear electro-optic effect enabling pure phase modulation [31]. This allows for the generation of ideal amplitude and phase modulated signals in an MZI configuration at moderate drive voltages [26], [29].

In this paper, we demonstrate generation of on-off-keying (OOK) and multi-level amplitude-shift-keying (ASK) signals using an SOH modulator. The device is operated at symbol rates up to 40 GBd with 2 levels (OOK, BPSK) and at a symbol rate of 28 GBd with 4 levels (4ASK) and 8 levels (8ASK). This is, to the best of our knowledge, the first demonstration of high-speed multilevel signal generation with a silicon based modulator. With this device we are able to generate a single polarization 8ASK signal at a data rate of up to 84 Gbit/s – the highest data rate so far generated by an electro-optic MZI modulator on silicon. In addition we demonstrate generation of Nyquist pulse-shaped BPSK and 4ASK signals at a data rate of 21 GBd.

Silicon-Organic Hybrid Modulator

Silicon-organic hybrid devices combine conventional silicon-on-insulator (SOI) waveguides with functional organic cladding materials [26], [30], [36], [43]. The cross-section of an SOH phase modulator is depicted in Fig. 5.21(a). It consists of two 240 nm wide and 220 nm high silicon rails separated by a 120 nm wide slot [27]. The waveguide is covered with an organic electro-optic cladding material. Field discontinuities at the slot sidewalls lead to strong confinement of the optical mode field within the electro-optic material in the slot. In this work we used the electro-optic polymer M3 of Gigoptix Inc. that has a nonlinear coefficient $r_{33} = 70$ pm/V if fully poled [82]. In addition, the rails are electrically connected to metal transmission lines by 45 nm thick *n*-doped silicon strips (stripload) such that the applied voltage drops predominantly across the 120 nm wide slot, resulting not only in a strong electric field, but also in a large overlap with the optical mode. A $U_\pi L$ product of 2 Vmm is expected for this waveguide geometry if the electro-optic cladding is fully poled. The individual devices which are reported here have $U_\pi L$ products of (6–10) Vmm corresponding to nonlinear coefficients of (23–14) pm/V, therefore poling can be improved. Other groups reached in-device nonlinear coefficients of up to 58 pm/V [99], underlining the potential of the approach to significantly reduce driving voltages.

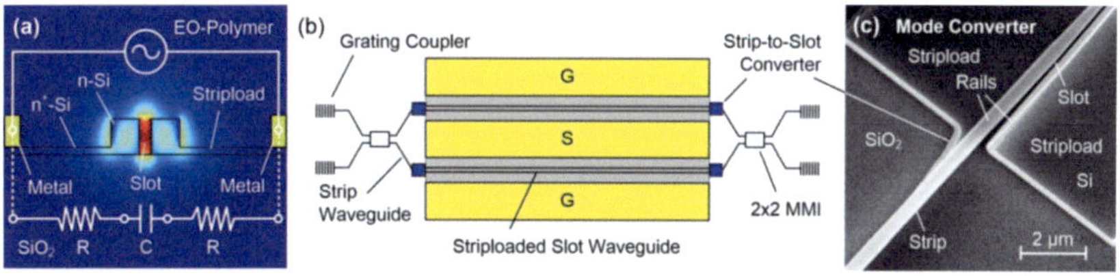

Fig. 5.21 Silicon-organic hybrid modulator. (a) Schematic and simulated mode of a phase modulator. The optical field is confined in the 120 nm wide slot. The two rails of the slot waveguide are electrically connected to metal electrodes by 45 nm high *n*-doped (As, $1 \ldots 4 \times 10^{17}\,\mathrm{cm}^{-3}$) silicon strips (stripload). In this way the applied potential drops over the 120 nm wide slot. This results in a high electric field in the slot and in a high overlap between optical and electrical mode. (b) Schematic of the MZI modulator. The structure consists of two phase modulators, driven by a single coplanar waveguide (CPW). 2×2 multimode interference couplers (MMI) are used as 3 dB power dividers. Grating couplers are used for fiber-chip coupling. (c) Scanning electron micrograph of a strip-to-slot converter used to couple access strip waveguides to the phase modulators.

The MZI modulator configuration is depicted in Fig. 5.21(b). It consists of a balanced MZI and two identical 1 mm long SOH phase modulators, Fig. 5.21(a). Two 2×2 multimode interference couplers (MMI) are used for the MZI. The silicon waveguide structures were fabricated on an SOI wafer with a 220 nm thick silicon layer and 2 µm thick buried oxide. Grating couplers are used to couple light from a singlemode fiber to the silicon chip [48]. Fully etched strip waveguides are used as on-chip access waveguides. Efficient coupling between strip and strip-loaded slot waveguides is ensured by using logarithmically tapered mode converters Fig. 5.21(c), as reported in [109]. A single coplanar transmission line is used to drive the push-pull MZI modulator. The two phase modulators have been poled with opposite polarities, referring to the signal electrode. The fiber-to-fiber loss is 20 dB with an on-chip loss of 9 dB in the C-Band. We measure a loss of 0.2 dB per MMI, 0.02 dB per strip-to-slot converter, 1 dB loss for the access waveguides and 7.5 dB loss in the 1 mm long active section. It should be noted that the loss of the striploaded slot waveguides in the active section increased only little after doping. Also propagation losses of the waveguides remain the same when a PMMA cladding is exchanged for the EO-polymer M3, and thus losses are not dominated by absorption of the cladding. Therefore the high propagation loss appears to be a fabrication issue. However, recently fabricated striploaded slot waveguides had a much lower loss below 1 dB/mm, indicating that the insertion loss of the devices can be significantly reduced by technology in future devices.

Data Transmission Setup

The experimental setup is shown in Fig. 5.22. Two different signal generators are set up to drive the modulator. A software-defined signal generator (FPGA+DAC, green) [110] is used to generate the 28 GBd *M*-ASK signals based on a pseudo-random bit sequence (PRBS) of length 2^{11}-1. For the generation of OOK and BPSK signals with symbol rates of up to 40 GBd a second pseudo-random pattern generator is used with a bit sequence length of 2^{31}-1. The electrical signal is amplified to have a voltage swing of up to 6 V_{pp} for driving the device under test (DUT) via a GSG-Picoprobe. A second Picoprobe with 50 Ω termination prevents reflections at the end of the device. Bias-Tees are used to adjust the operating point ($\varphi_{\mathrm{Bias}} \propto U_{\mathrm{Bias}}$) of the MZI modulator. In order to increase the conductivity of the silicon stripload, Fig. 5.21(a), a gate voltage has been applied between substrate and stripload, see Fig. 5.22. A detailed description of the effect of the gate voltage can be found in reference

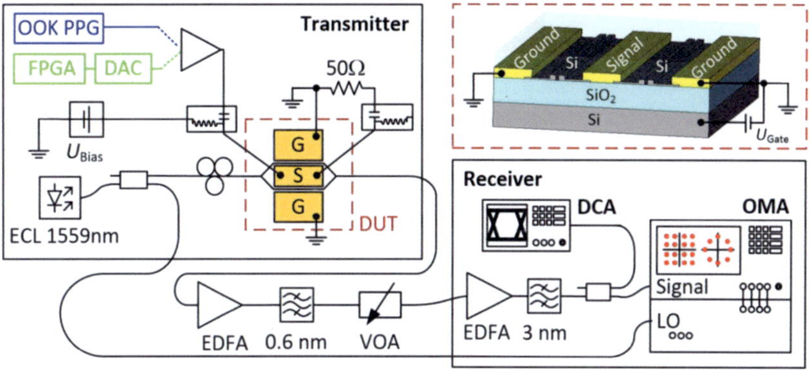

Fig. 5.22 Experimental setup. Two different signal generators are used. For generation of on-off-keying (OOK) and binary-phase-shift-keying (BPSK) with symbol rates up to 40 GBd a pseudo-random bit pattern generator (blue) is used. A second (software defined) signal generator (FPGA+DAC, green) is used to generate the *M*-ASK signals with data rates up to 28 GBd. The signal is fed into the coplanar waveguide (CPW) of the modulator using a GSG Picoprobe and terminated by a second Picoprobe at the end of the modulator. CW light is coupled into the device after adjusting polarization. The same CW light serves as local oscillator (LO) for the coherent detector. After amplification the signal is detected by a digital communications analyzer (DCA) and by a coherent receiver (optical modulation analyzer – OMA).

[31]. Light from an external cavity laser (ECL) at a wavelength of 1559 nm is coupled to the DUT. The input power is 8.5 dBm and the polarization is adjusted to be quasi-TE on the chip. The same light source serves as local oscillator for the coherent receiver (optical modulation analyzer – OMA, Agilent Technologies N4391A). After the DUT, the signal is amplified using two cascaded erbium doped fiber amplifiers (EDFA) and subsequently detected by a digital communications analyzer (DCA) and by a coherent receiver.

Experimental Results

In this work SOH modulators are used to generate NRZ-OOK signals with bit rates up to 40 Gbit/s, *M*-ASK signals with bit rates up to 84 Gbit/s and *M*-ASK Nyquist pulse-shaped signals at a symbol rate of 21 GBd. OOK experiments were performed with an MZI modulator fabricated by 193 nm deep UV lithography at IMEC, Belgium with CMOS-like copper metallization and $U_\pi = 6\text{ V}$, while the coherent experiments were carried out using a second MZI modulator with identical design dimensions, fabricated by electron beam lithography at AMO, Germany, that has aluminum electrodes and $U_\pi = 10\text{ V}$.

5.6.1 Generation of On-Off-Keying Signals

In the first experiment the pseudo-random bit pattern generator (PPG, Fig. 5.22, blue) is used to generate the electrical drive signal. The signal is amplified to have a voltage swing of 6 V_{pp}. Fig. 5.23 summarizes the measured eye diagrams and bit error ratios (BER) at various data rates. At 25 Gbit/s and 40 Gbit/s the extinction ratios (ER) are 13.3 dB and 11.3 dB. No bit errors have been detected up to a data rate of 35 Gbit/s in the chosen measurement time indicating a BER below 1×10^{-12}. At 40 Gbit/s a BER of 1×10^{-11} is measured.

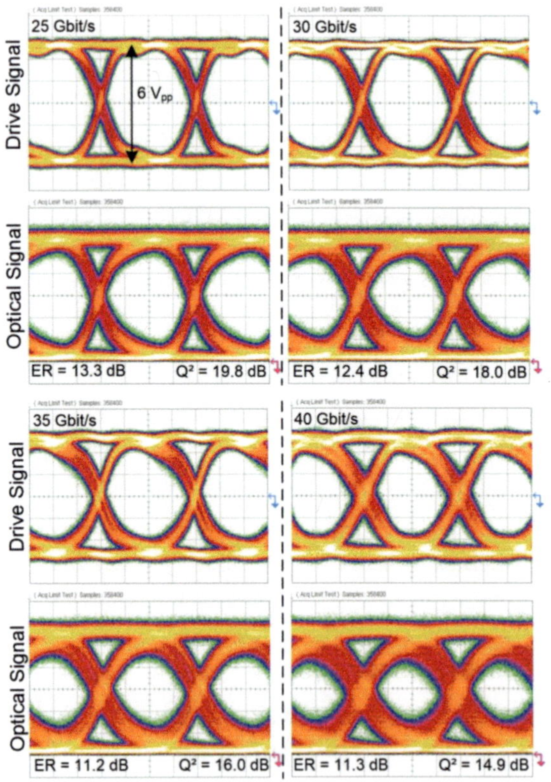

Fig. 5.23 On-off-keying experiment. Recorded eye diagrams at various data rates. Measured extinction ratios (ER) and Q^2-factors are depicted in the figure. The ER is above 11 dB at all data rates. The bit error ratio (BER) was measured. In the chosen measurement time no bit errors have been received for data rates up to 35 Gbit/s. At 40 Gbit/s the measured BER is 1×10^{-11}. A drive voltage of 6 V_{pp}, a gate field of 150 V/µm and a PRBS of length 2^{31}-1 is used.

5.6.2 Generation of *M*-ASK Signals

Single-polarization BPSK signals at symbol rates of 25 Gbd to 40 GBd are generated, Fig. 5.24(a-c), using the same pattern generator as before (Fig. 5.22, blue) and adjusting the working point to the Null point. The optical signal was detected by the OMA and recorded with 80 GSamples/s. Digital signal processing and signal analysis is performed offline. An equalizer with a filter length of 55 symbols is used to compensate bandwidth limitations of the transmitter electronics, the OMA and the SOH-modulator. No bit errors have been measured in a set of 10 million recorded samples at these symbol rates although a degradation of the error vector magnitude (EVM_m [111], [112]) from 7.2% (25 Gbit/s) to 12.1% (40 Gbit/s) is observed.

Next BPSK, 4ASK and 8ASK signals are generated and received at a symbol rate of 28 Gbd by using a software-defined signal generator (Fig. 5.22, green), that consists of a field-programmable gate array (FPGA) and a digital-to-analog converter (DAC) [110]. The length of the bit sequence was 2^{11}-1. For operating the modulator in the linear regime the drive voltage is reduced to 5.6 V_{pp}. The experimental results are depicted in Fig. 5.24(d-f). An EVM_m of 10.2% is measured for BPSK, Fig. 5.24(d). No errors are measured in a set of 3.5 million recorded bits. In the case of 4ASK, Fig. 5.24(e), an EVM_m of 9.9%

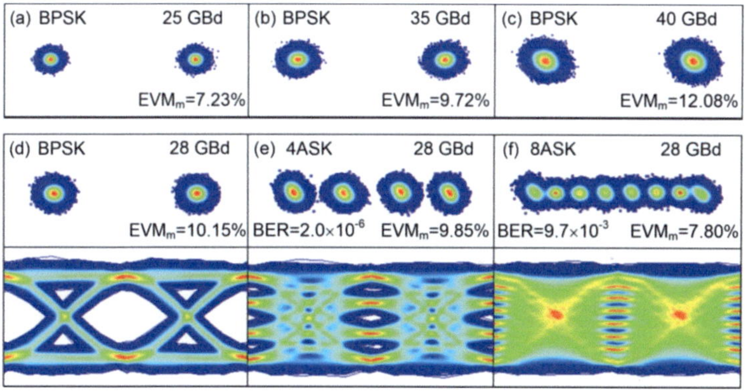

Fig. 5.24 Constellation diagrams and eye-diagrams of the *M*-ASK signals. (a)-(c) BPSK signals at symbol rates of 25 GBd (EVM_m 7.2%), 35 GBd (EVM_m 9.7%) and 40 GBd (EVM_m 12.1%). No bit errors were detected. (d)-(f) *M*-ASK signals at a symbol rate of 28 GBd. (d) BPSK. The EVM_m is 10.2%. No bit errors were received. (e) 4ASK. The EVM_m is 9.9% with a BER of 2×10^{-6}. (f) 8ASK. The EVM_m is 7.8% and the BER was 9.7×10^{-3}. A drive voltage of 5.6 V_{pp}, a gate field of 150 V/µm and a PRBS of length 2^{11}-1 was used.

and a BER of 2×10^{-6} is measured, well below the threshold of hard-decision FEC of 3×10^{-3}. Next an 8ASK signal is generated, Fig. 5.24(f), resulting in a gross data rate of 84 Gbit/s. An EVM_m of 7.8% and a BER of 9.7×10^{-3} is measured, which is below the threshold of soft-decision FEC [113] (BER = 1.9×10^{-2}) resulting in a net data rate of 67.2 Gbit/s (25% overhead). Taking also into account the 50 Ω termination of the transmission line we estimate the energy consumption of the modulator to be 800 fJ/bit for the generation of 28 GBd 8ASK signals. We believe that improved poling procedures and device optimizations will allow to considerably increase the performance of the modulator: In the current device, an r_{33} coefficient of 14 pm/V is obtained, measured at 5 GHz – a factor of 5 below the values obtained in polymer waveguides of the same material [82].

5.6.3 Generation of *M*-ASK Nyquist Pulse-Shaped Signals

Nyquist pulse-shaping has been found to be a good candidate to increase spectral efficiency of optical communication systems [114], [115]. While in wavelength division multiplex (WDM) systems the spectrum of each carrier is usually infinitely broad, a finite rectangular spectrum with the bandwidth of the modulation frequency can be achieved by Nyquist pulse-shaping, making guard intervals no longer necessary [116]. To do so each bit is represented by a sinc-function in time-domain, infinitely expanded in the ideal case. The Fourier transform of the sinc-function in time-domain is a rectangular spec-

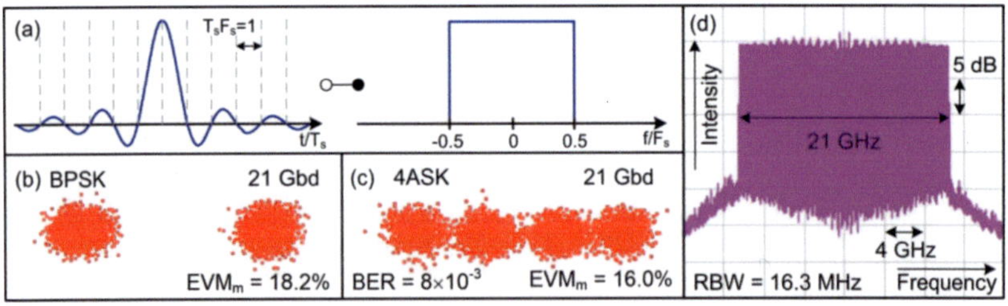

Fig. 5.25 Nyquist pulse shaping. (a) Nyquist pulse-shaped bit. The envelope of each bit is a sinc-function. The zero points of the sinc-functions are allocated between the bit slots T_S. The spectrum (Fourier transform) of a Nyquist pulse-shaped signal is a rectangular function. This allows for highly spectrally efficient WDM systems. (b,c) Constellation diagrams of Nyquist BPSK and 4ASK signals at a symbol rate of 21 GBd. For BPSK an EVM_m of 18.2% and no bit errors have been measured. For 4ASK an EVM_m of 16% is measured. The measured BER is 8×10^{-3}. (c) Spectrum of the Nyquist pulse-shaped signal. The rectangular shape of the spectrum is confirmed and an extinction ratio of 20 dB is measured.

trum in frequency domain, Fig. 5.25(a). The zero positions of the sinc-function are allocated in between two bit slots T_S. A Nyquist pulse-shaped PRBS of length 2^9-1 is loaded to the software-defined signal generator. At a clock frequency of 28 GHz at the DAC the signal is oversampled by a factor 4/3, resulting in a symbol rate of 21 GBd [117]. The constellation diagrams for BPSK and 4ASK are depicted in Fig. 5.25(b,c). EVM_m of 18.2% and 16% are measured respectively. Fig. 5.25(d) shows the optical spectrum of the 4ASK signal. The expected rectangular shape is confirmed. The consumed optical bandwidth is 21 GHz and the extinction ratio is 20 dB.

Conclusion

We experimentally demonstrate, for the first time, generation of multi-level signals in SOH electro-optic modulators. Our device is operated at a symbol rate of 28 GBd with up to 8 symbols (8ASK), resulting in a data rate of 84 Gbit/s. For BPSK and OOK a symbol rate of 40 GBd is successfully demonstrated. This is, to the best of our knowledge, the highest symbol rate generated by an SOH Mach-Zehnder Interferometer modulator, enabling the highest data rate generated by a single electro-optic MZI modulator on silicon. In addition, Nyquist pulse-shaping has been performed for the first time by using a silicon modulator. We believe that our results are an important step towards 16QAM and 64QAM modulation using silicon devices in near future.

[End of Paper J5]

5.7 Modulation at Low Energy Consumption

Energy consumption of a modulator is predominantly determined by the voltage required to drive the modulator. According to Eq. (2.39) the π-voltage of the device can be minimized by optimization of the device geometry and by optimizing the material parameters. In the following, two exemplary realizations of SOH MZM with low energy consumption are demonstrated. In the first example, Section 5.7.1, low energy consumption was realized by optimizing device geometry, choosing a slot width of 80 nm and a device length of 1.5 mm. The second example, Section 5.7.2, was realized by using an EO organic cladding with extraordinary high EO coefficient of 180 pm/V.

5.7.1 Reducing Energy Consumption by Modulator Design

This section was published in a scientific journal [J7]:

[Begin of Paper J7]

Low Power Mach-Zehnder Modulator in Silicon-Organic Hybrid Technology

IEEE Photonic Technology Letters; Volume 25; Issue 13; pp. 1226-1229; July 2013
DOI: 10.1109/LPT. 2013.2260858

R. Palmer,[1] L. Alloatti,[1] D. Korn,[1] P. C. Schindler,[1] M. Baier,[1] J. Bolten,[2] T. Wahlbrink,[2] M. Waldow,[2] R. Dinu,[3] W. Freude,[1] C. Koos,[1] J. Leuthold,[1,6]

[1]*Karlsruhe Institute of Technology (KIT), Engesserstr. 5, 76131 Karlsruhe, Germany*

[2]*AMO GmbH, Aachen, Germany*

[3]*GigOptix Inc., Switzerland and GigOptix Bothell, Washington, USA*

We report on a silicon-organic hybrid (SOH) modulator based on a Mach-Zehnder interferometer (MZI) operating at 10 Gbit/s with an energy consumption of 320 fJ/bit. The device consists of a striploaded slot waveguide covered with an electro-optic polymer cladding. The MZI modulator is poled to be driven in push-pull operation by a single coplanar RF line. Our nonlinear coefficient r_{33} = 15 pm/V in combination with an 80 nm narrow slot enable RF peak-to-peak drive voltages as low as 800 mV to suffice for an extinction ration of 4.4 dB for a 1.5 mm long modulator.

Introduction

Silicon electro-optic modulators are key elements in telecommunications, data-communications and optical interconnects. So far, modulators were demonstrated based on free carrier depletion [103], [118], [119] or injection [18], [102] in diode structures or MOS (metal-oxide-semiconductor) structures [120]. Also the Pockels effect was exploited in silicon slot waveguides [27] using an electro-optic organic cladding [26], [29]–[31], [36] or in strained silicon rib-waveguides [15], [16]. A major challenge is to keep RF modulator voltages compatible with CMOS driver electronics, removing the need of power consuming RF amplifiers. Other important modulator parameters are device footprint, electrical and optical bandwidth, π-voltage-length product $U_\pi L$, extinction ratio, and energy consumption [17]. Resonant structures such as ring resonators [101], [121], [102], [122] can be operated with low drive voltages, and have a small foot-print as well, but their performance depends strongly on wavelength and operation temperature. For stabilization, energy consuming feedback loops are necessary, thereby limiting the field of application. In contrast, modulators based on Mach-Zehnder interferometers (MZI) have only weak temperature and wavelength dependency, but a larger footprint and a rather large operating voltage. Devices based on carrier depletion prove to be inferior to injection based devices in terms of length and operation voltage, but are superior in terms of bandwidth and energy-consumption [17]. Therefore, depletion based devices are usually preferred for high-speed applications. Voltage-length products of $U_\pi L$ are typically larger than 10 Vmm for non-resonant structures, and extinction ratios are in the order of 3…10 dB [17]. For reducing the drive voltages of depletion MZI modulators, the length of the phase modulator region can be extended to several millimeters. This way an MZI depletion modulator was reported featuring a length of 5 mm, a bias dependent $U_\pi L$ of 10…50 Vmm and 4.2 dB extinction ratio at 20 Gbit/s for a drive voltage of 1 V_{pp} [23].

Here we report of an MZI modulator based on the silicon-organic hybrid (SOH) approach [26]. Our MZI modulator has a short length of 1.5 mm, a voltage length product $U_\pi L$ of 2.7 Vmm and an extinction ratio of 4.4 dB measured at 10 Gbit/s for a drive voltage of 800 mV_{pp}. The low drive voltage will enable photonic-electronic integration without RF amplifiers on a small footprint.

Silicon-Organic Hybrid Modulator

Silicon-organic hybrid devices combine conventional silicon-on-insulator (SOI) waveguides with functional organic cladding materials. The cross-section of an SOH phase modulator is depicted in Fig. 5.26(a). It consists of two 240 nm wide silicon rails separated by an 80 nm wide slot. The dimensions have been verified with a scanning electron microscope, Fig. 5.26(b). The waveguide is covered by an organic electro-optic cladding. In this work we used the electro-optic polymer M3 of GigOptix Inc. [82]. Discontinuities of the dominant (horizontal) electric field component at the slot sidewalls lead to a strong light confinement of the quasi-TE mode within the slot. The two rails of the slot waveguide are connected to a metal transmission line by 45 nm thick *n*-doped (As, $1.5{\cdot}10^{17}\mathrm{cm}^{-3}$) silicon strips (stripload). This way the applied voltage drops entirely across the 80 nm wide slot. This results in a strong electric field that has a large overlap with the optical mode [29], [123]. The conductivity of the striploads can be further enhanced by applying a gate-voltage U_{Gate} between silicon substrate and stripload, as depicted in Fig. 5.26(a). A more detailed description of the "gate-functionality" can be found in reference [31]. The schematic of the MZI modulator with multimode interference (MMI) couplers is depicted in Fig. 5.26(c). It consists of a balanced MZI and two identical 1.5 mm long SOH phase-modulators, Fig. 5.26(a). The silicon waveguides were fabricated on an SOI wafer with a 220 nm thick device layer and a 3 µm thick buried oxide using a fabrication process similar to the one described in [109]. Grating couplers couple light from a singlemode fiber to the silicon chip [48]. Strip waveguides are used as on-chip access waveguides. Efficient coupling between strip and striploaded slot waveguides is ensured by using logarithmically tapered mode converters [109]. A coplanar transmission line, consisting of 450 nm thick aluminum lanes, is used to drive the single-drive push-pull MZI modulator. The two phase modulators were poled at elevated temperatures with opposite polarities by applying the poling voltage across the ground electrodes of the coplanar transmission line as depicted in Fig. 5.26(d). In this way exactly half of the applied poling voltage drops across each silicon slot. For poling, the sample is heated to 168 °C, and a poling voltage of 44 V (22 V per phase-modulator) is applied. At room temperature, the poling current is in the nA-region indicating that both rails are well insulated.

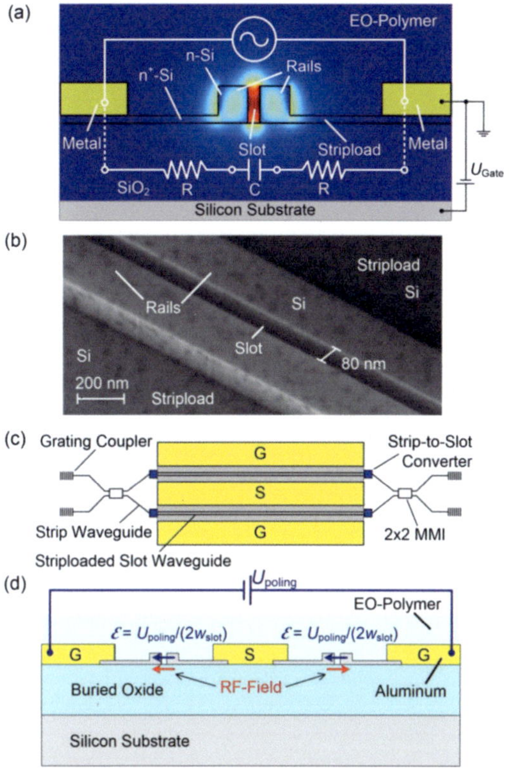

Fig. 5.26 Silicon-organic hybrid MZM. (a) Schematic and simulation of the quasi-TE polarized mode of the phase modulator. The dominant electric field component is parallel to the substrate. Field discontinuities at the slot boundaries lead to a strong light confinement within the 80 nm wide slot. The two rails of the slot waveguide are electrically connected to metal electrodes by 45 nm thick *n*-doped (As, $1.5 \cdot 10^{17}\text{cm}^{-3}$) silicon strips (stripload). In this way the applied potential drops across the 80 nm wide slot. This results in a high electric field in the slot and in a high overlap between optical and electrical mode. (b) SEM image of the striploaded slot waveguide. The measured slot width is 80 nm. The dark spots are residues of the polymer cladding. (c) Schematic of the device. The MZI consists of two phase modulators as described in (a) operated in push-pull configuration by a single coplanar line (GSG: ground – signal – ground) and two 2×2 multimode interference couplers (MMI). Light is coupled to the chip using grating couplers. (d) Cross-section of the modulator. Schematic illustration of how the poling voltage is applied to achieve push-pull operation. The voltage source is connected to the two ground electrodes for poling. In that way half of the poling voltage drops across each slot and the global orientation of the chromophore dipoles in both phase modulators is identical.

Characterization

In this section static and dynamic properties of the modulator are investigated first. Then, on-off keying at a data rate of 10 Gbit/s is demonstrated.

A. Modulator Properties

First, static properties of the modulator are investigated, Fig. 5.27(a). A gate field of 66 V/µm is applied between the stripload and the substrate for increasing the conductivity of the striploads, Fig. 5.26(a). In Fig. 5.27(a) we depict the MZI characteristic as a function of the bias voltage. Depending on the operat-

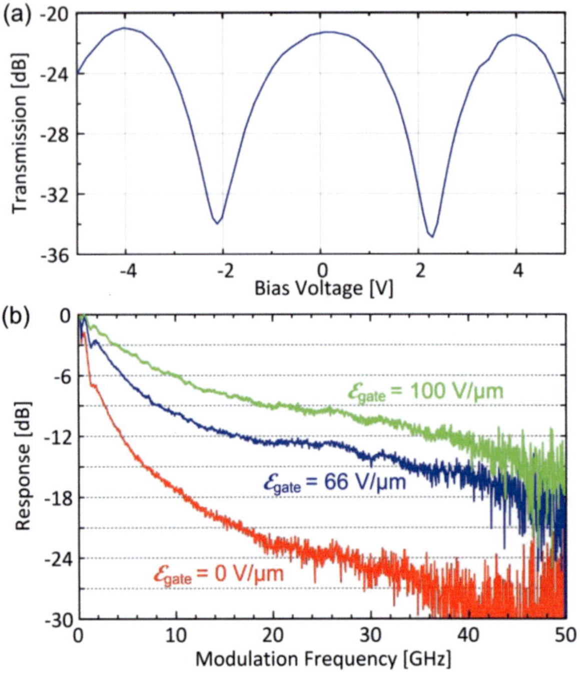

Fig. 5.27 Static and dynamic properties of the MZI modulator. (a) DC operation of the modulator with a gate field of 66 V/µm. Dependent on the operating point, a U_π between 1.8 V and 2.2 V is measured. (b) Electrical-optical-electrical response of the MZI modulator measured for 3 different gate fields. The gate field, see Fig. 5.26(a) has a strong impact on the modulator bandwidth indicating that the doping concentration in the silicon striploads is not sufficient for high speed operation. The gate field increases the conductivity of the striploads. At a field of 100 V/µm the 6 dB bandwidth is 10 GHz. The bandwidth could be further enhanced by increasing the gate field. Dielectric breakdown of the SiO_2-layer occurs at 900 V/µm [124].

ing point (e. g. 0 V, ±4 V) the U_π voltage is either 2.2 V or 1.8 V. From these data, and with the waveguide dimensions $w_{slot} = 80$ nm, $w_{rail} = 240$ nm, $h_{stripload} = 45$ nm we calculate a nonlinear coefficient r_{33} of 15 pm/V, while the theoretically achievable value is 70 pm/V [82]. Next the electrical-optical-electrical (EOE) frequency response of the modulator is measured and depicted in Fig. 5.27(b). The measurement is performed with an electrical vector network analyzer, a CW laser, the device under test (DUT) and a high-speed photodetector. If no gate field is applied, we observe that the bandwidth of the modulator is below 2 GHz, indicating that the conductivity of the doped strip-loads is not sufficient for high-speed operation for the chosen doping concentration [29]. When applying the gate field the stripload conductivity increases, boosting the 6 dB-bandwidth to 10 GHz for a field of 100 V/µm. The bandwidth could be further enhanced by increasing the gate field. An SiO_2 layer can stand voltages up to 900 V/µm [124].

The fiber-to-fiber loss of the device is 21 dB with an on-chip loss of approximately 9 dB.

B. On-Off Keying Experiment

The modulator is tested in a system experiment as depicted in Fig. 5.28. A gate field of 66 V/µm is applied. A 10 Gbit/s signal is generated by a pseudorandom bit pattern generator (PPG), amplified and fed to the RF electrodes of the modulator using GSG picoprobes. A bias voltage is applied via the same GSG

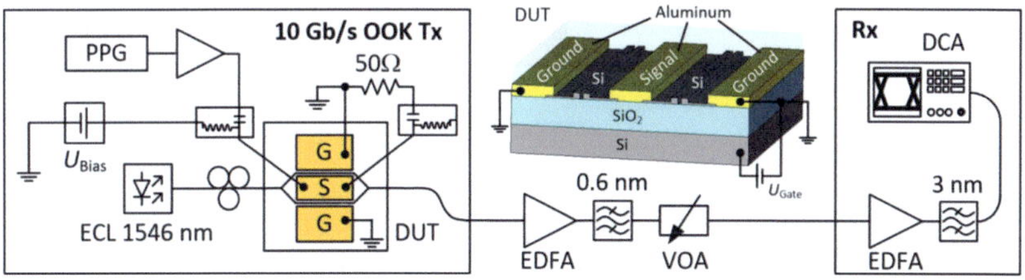

Fig. 5.28 Experimental setup. A 10 Gbit/s pseudorandom bit pattern generator (PPG) is used to drive the MZM. The signal is amplified and fed into the modulator using GSG-picoprobes. A bias voltage is applied over the same GSG electrodes by using a bias-T. A second picoprobe at the end of the modulator is used to terminate the electrical signal. Light from an external cavity laser (ECL) with controlled polarization at a wavelength of 1546 nm is coupled to the silicon chip using grating couplers. The signal is subsequently amplified and received by a digital communications analyzer (DCA). The inlet depicts a schematic of the MZM. A gate voltage U_{Gate} is applied between silicon substrate and transmission line to increase the conductivity of the stripload [31].

electrodes using a bias-T. A second picoprobe at the end of the modulator terminates the electrical signal. Light at a wavelength of 1546 nm is coupled to the modulator from an external cavity laser (ECL) using grating couplers. The polarization is set to be quasi-TE polarization on the silicon chip. The modulated light leaving the DUT is amplified and received by a digital communications analyzer (DAC).

Recorded eye diagrams for various drive voltages are depicted in Fig. 5.29. Wide and open eyes with an extinction ratio of (4…11) dB can be seen for peak-to-peak drive voltages as small as 800 mV_{pp} (400 mV_p), see Fig. 5.29(a). Increasing the drive voltage to 2.5 V_{pp} results in an optimum eye diagram with an extinction ratio of 11 dB and a high signal quality, see Fig. 5.29(c) (Q^2-factor of 22 dB). Increasing the drive voltage to 3 V_{pp} and 4 V_{pp} leads to an over-modulation, Fig. 5.29(d,e).

In the current operation mode the drive signal is terminated in a 50 Ω resistor at the end of the modulator. Adding the energy dissipation in the resistor to the energy calculation results in an energy consumption of 320 fJ/bit for a drive voltage of 800 mV_{pp}. This value is among the lowest so far reported for a silicon MZI modulator at this data rate [17], [23]. For a similar but shorter modulator of 1 mm length we verified that such devices can be operated without termination at data rates of up to 25 Gbit/s. Reflections are not an issue due

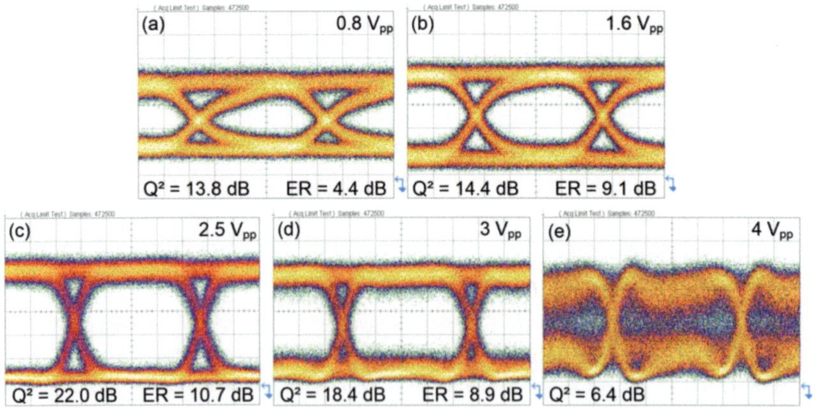

Fig. 5.29 10 Gbit/s NRZ-OOK eye-diagrams. At this data rate we find the π-voltage of the modulator to be 2.5 V_{pp}. Increasing the drive voltage to 3 V_{pp} or above leads to over-modulation. Open eye diagrams are also achieved when the drive voltage is reduced to 1.6 V_{pp} and 800 mV_{pp}. The measured Q^2-factors and extinction ratios (ER) are noted in the figure. 200 μV/div.

to the short length of the modulator, which is much smaller than the wavelength of the RF signal. Though the modulator that is used in this work is 50% longer it is reasonable to assume that it can be operated without termination at the reduced data rate of 10 Gbit/s. With an ideal voltage source and without a 50 Ω termination the energy consumption would be dominated by capacitive loading. For an on-chip drive voltage of 800 mV_{pp} and a calculated slot capacitance of $C = 189$ fF we then can estimate the energy consumption of the modulator per bit to be as low as $(1/4)(2C)V_{pp}^2 = 60$ fJ.

Conclusion and Outlook

We have demonstrated a 1.5 mm long push-pull silicon-organic hybrid MZI modulator with a U_π of 1.8 V at DC and 2.5 V at a data rate of 10 Gbit/s, which corresponds to a polymer nonlinearity of $r_{33} = 15$ pm/V. Open eye diagrams and an extinction ratio of (4…11) dB are demonstrated for drive voltages as low as 800 mV_{pp} making electrical RF amplifiers dispensable. Improvements are possible concerning the poling procedure. With a fully poled structure the limit for U_π would be around 390 mV for this Mach-Zehnder structure and the chosen nonlinear polymer. The result forecasts ultra-low energy consuming electro-optic interconnects with silicon modulators directly driven by on-chip CMOS digital-to-analog converters.

[End of Paper J7]

5.7.2 Reducing Energy Consumption by Organic Material

This section has been published in a scientific journal [J15] and in part in the proceedings of a scientific conference [C28].

[Begin of Paper J15]

Femtojoule Electro-Optic Modulation Using a Silicon-Organic Hybrid Device

Nature - Light: Science & Applications; Volume 4; Issue 2; p. e255; February 2015
DOI:10.1038/lsa.2015.28

S. Koeber[1†], R. Palmer[1†], M. Lauermann[1], W. Heni[1,2], D. L. Elder[3], D. Korn[1], M. Woessner[1], L. Alloatti[1,4], S. Koenig[1,5], P. C. Schindler[1,5], H. Yu[6,7], W. Bogaerts[6], L.R. Dalton[3], W. Freude[1], J. Leuthold[1,2], C. Koos[1]

[1]*Institute of Photonics and Quantum Electronics (IPQ) and Institute of Microstructure Technology (IMT), Karlsruhe Institute of Technology (KIT), 76131 Karlsruhe, Germany*

[2]*Now with: Institute of Electromagnetic Fields, Swiss Federal Institute of Technology (ETH), Zurich, Switzerland*

[3]*University of Washington, Department of Chemistry, Seattle, WA 98195-1700, United States*

[4]*Now with: Massachusetts Institute of Technology, Research Lab of Electronics (RLE), Cambridge, MA 02139, United States*

[5]*Now with: Infinera Corporation, Sunnyvale, CA, USA*

[6]*Ghent University – IMEC, Photonics Research Group, Department of Information Technology, Gent, Belgium*

[7]*Now with: Department of Information Science and Electronic Engineering, Zhejiang University, Hangzhou 310027, China*

[†]*these authors contributed equally to the work*

Abstract

Energy-efficient electro-optic modulators are at the heart of short-reach optical interconnects, and silicon photonics is considered the leading technology for realizing such devices. However, the performance of all-silicon devices is limited by intrinsic material properties. In particular, the absence of linear electro-optic effects in silicon renders the integration of energy-efficient photonic-electronic interfaces challenging. Silicon-organic hybrid (SOH) integration can overcome these limitations by combining nano-photonic silicon waveguides with organic cladding materials, thereby offering the prospect of designing optical properties by molecular engineering. In this paper, we demonstrate an SOH Mach-Zehnder modulator with unprecedented efficiency: The 1 mm-long device consumes only 0.7 fJ/bit to generate a 12.5 Gbit/s data stream with a bit-error ratio below the threshold for hard-decision forward-error correction (FEC). This power consumption represents the lowest value demonstrated for a non-resonant Mach-Zehnder modulator in any material system. It is enabled by a novel class of organic electro-optic materials that are designed for high chromophore density and enhanced molecular orientation. The device features an electro-optic coefficient of $r_{33} \approx 180$ pm/V and can be operated at data rates of up to 40 Gbit/s.

Introduction

Optical interconnects are the most promising option to overcome transmission bottlenecks in data centres and high-performance computers, and energy consumption is one of the most important parameters of the associated photonic-electronic interfaces. Targeted figures are tens of femtojoule per bit for transmitters in off-chip connections, and a few femtojoule per bit for on-chip links [125]. Key requirements are low drive voltages that can be provided by standard CMOS circuitry without further amplification. Apart from energy efficiency, modulators need to provide fast electro-optic (EO) response along with large optical operation bandwidth to ensure high-speed transmission and flexibility in wavelength-division multiplexing (WDM) systems. Moreover, dense integration is essential, calling for a small device footprint.

Silicon photonics is currently the most promising technology to realize such devices, leveraging mature high-yield CMOS processing and offering the potential of photonic-electronic co-integration on large-area silicon wafers. However, second-order nonlinearities are absent in bulk silicon due to inversion symmetry of the crystal lattice. Hence, current silicon-based modulators [17]-[120] have to rely on free-carrier depletion or injection in p-n, p-i-n, or metal-oxide-semiconductor (MOS) structures. This leads to various trade-offs when realizing fast and energy-efficient devices with small footprint: While carrier injection in forward-biased p-i-n-structures enables compact modulators with voltage-length products as low [126] as $U_\pi L = 0.36$ Vmm, free-carrier lifetime currently limits the modulation speed to 25 Gbit s^{-1} and requires strong pre-emphasis of the drive signal [127]. Moreover, energy efficiency of these devices is intrinsically limited to the pJ/bit-range due to the permanent injection current flowing through the diode section. In contrast, carrier depletion in reverse-biased p-n junctions enables negligible bias currents and symbol rates [20] of up to 50 GBd, but typical voltage-length products are beyond 10 Vmm and thus much larger than those of injection-type devices. For a silicon-based Mach-Zehnder modulator (MZM), the lowest reported energy consumption amounts to 200 fJ/bit, achieved in a 5 mm-long depletion-type device [23]. Modulation energies and device footprint can be significantly reduced by using resonant structures such as microrings, microdiscs, or photonic-crystal waveguides [17], [24]. For resonant silicon-based modulators, the lowest energy consumption reported to date amounts to 3 fJ/bit and has been achieved with a microdisc device [24] operated at a data rate of 12.5 Gbit/s. However, the optical bandwidth of resonant devices is inherently limited, and resonance wavelengths are often subject to strong temperature-induced drifts.

Our work aims at overcoming the intrinsic limitations of all-silicon devices by combining conventional silicon-on-insulator (SOI) slot waveguides [128] with organic cladding materials in a hybrid approach. The concept of silicon-organic hybrid (SOH) integration [46], [129] leverages both the benefits of large-scale standardized CMOS processing and the wealth of optical properties provided by theory-guided molecular design of organic materials [97]. In particular, the SOH concept lends itself to highly efficient EO modulators that feature a large modulation bandwidth, a broad range of operation wavelengths, and a small device footprint [70], [130], [131]. First SOH EO devices have been demonstrated using guest-host systems of polymer materials that are doped with EO molecules, so-called chromophores [31], [35], [109], [132]–[134]. When used as thick layers between transparent electrodes, such materials exhibit high EO coefficients [69] of up to 118 pm/V. However, for nano-photonic SOH modulators, in-device EO coefficients are found to be much lower with typical values in the range [31], [36], [70], [77], [81], [99] of 20 pm/V to 60 pm/V, leading to energy consumptions [134] of 300 fJ/bit or more. These figures are comparable to those of all-silicon devices, but still far from exploiting the full potential of the SOH approach.

Here, we demonstrate that an ultra-low energy consumption of down to 0.7 fJ/bit can be achieved on the silicon integration platform even with non-resonant devices. In a proof-of-concept experiment, a 1 mm-long SOH Mach-Zehnder modulator is operated with drive voltages as small as 80 mV_{pp} at a data rate of 12.5 Gbit/s with a bit-error ratio below the threshold for hard-decision forward-error correction (FEC) [72]. This demonstration is enabled by combining an optimized SOH slot-waveguide modulator with a novel class of organic EO materials, which consists of pure chromophores instead of a chromophore-polymer guest-host system. The modulator exhibits an in-device EO coefficient of 180 pm/V, and enables modulation with peak-to-peak drive voltages of down to 80 mV at 12.5 Gbit/s. The devices support data rates of up to 40 Gbit/s with an extinction ratio exceeding 10 dB. At DC, the $U_{\pi}L$ product is as low as 0.5 Vmm.

MATERIALS AND METHODS

Silicon-organic hybrid electro-optic modulator

A schematic and a cross-sectional view of an SOH MZM is displayed in Fig. 5.30(a) and (b), respectively. The device consists of two 1 mm-long SOH phase modulators that are driven in push-pull configuration by a single coplanar transmission line with a ground-signal-ground (GSG) configuration. A cross-sectional view of a single SOH phase modulator is depicted in Fig. 5.30(c) along with the electric field profile of the fundamental quasi-TE mode. The phase modulator section consists of an SOI slot waveguide that comprises two silicon rails separated by a 160 nm wide slot. The waveguide is covered and the slot is filled with an organic EO material that is deposited by spin-coating. The rails of the slot waveguide are electrically connected to a copper RF transmission line by thin *n*-doped silicon slabs and tungsten vias. A voltage applied to the transmission line thus drops across the narrow slot, resulting in a strong modulation field that has a large overlap with the guided optical mode.

For high-modulation efficiency, an organic material featuring a large macroscopic EO coefficient r_{33} is essential. Besides the optical properties of the individual chromophore molecule, the macroscopic EO coefficient depends on the orientation of the chromophore ensemble with respect to the modulating RF field. After deposition of the organic material, the EO coefficient is zero ($r_{33} = 0$) due to the random orientation of the dipolar chromophores. For achieving a non-centrosymmetric orientation (acentric order), the material must be poled. This is achieved by heating the organic material close to its glass-transition temperature T_g while applying an external electric field to align the dipolar chromophore molecules. The material is subsequently cooled back to room temperature while maintaining the poling field, thereby conserving the acentric order of the chromophores and the macroscopic EO coefficient of the material. For poling of our device, we apply the poling voltage U_{poling} between the two floating ground electrodes of the RF transmission line [134], as depicted in Fig. 5.30(b). This way, half of the poling voltage drops across

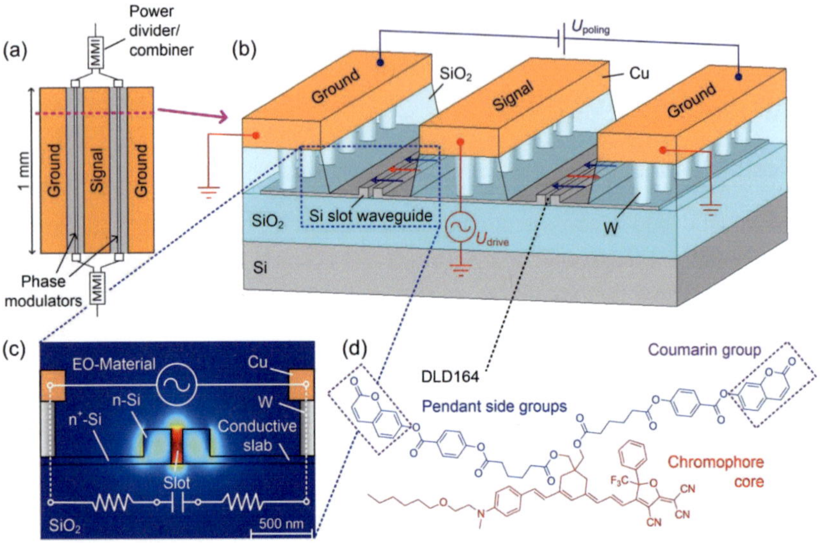

Fig. 5.30 Schematic of an SOH Mach-Zehnder modulator. (a) Two 1 mm-long silicon slot waveguides act as phase modulators and are driven in push-pull mode by a ground-signal-ground coplanar transmission line. Further details of the photonic integrated circuit are explained in the Supplementary Information. (b) Cross-section of the MZM. The transmission line consists of 600 nm-thick copper (Cu) electrodes connected to the silicon slot waveguides by 900 nm-high tungsten (W) vias. The organic cladding (not shown) is deposited into the slot region of the silicon-on-insulator (SOI) waveguide and is then poled at an elevated temperature by applying a poling voltage U_{poling} between the floating ground electrodes of the coplanar transmission line. This leads to a strong poling field $\mathcal{E}_{\text{poling}} = U_{\text{poling}} / 2w_{\text{slot}}$ in each slot that aligns the dipolar chromophores as indicated by dark blue arrows. The alignment is antisymmetric with respect to the RF modulation field marked by red arrows, resulting in a push-pull operation of the MZM. (c) Schematic and simulated quasi-TE optical mode of a slot waveguide that is filled with an EO organic material. The two rails of the slot waveguide are connected to the transmission line by 60 nm-thick n-doped silicon slabs. The optical mode and the RF modulation field are both well confined to the slot region. This leads to strong EO interactions and therefore to a high modulation efficiency. (d) Chemical structure of the EO cladding material [69] DLD164. The molecule consists of a "ring-locked" EO chromophore core (red), and two coumarin-based pendant side chains (blue).

each silicon slot, resulting in an orientation of the dipolar chromophores (dark blue arrows), which is parallel to the RF modulation field (red arrows) in the left half of the GSG transmission line, and anti-parallel in the right half. This leads to a positive phase shift in the left arm of the MZM, and to a negative phase shift in the right arm, thereby resulting in push-pull operation of the device.

The optical fibre-to-fibre insertion loss of the modulator amounts to 16.5 dB, dominated by fibre-to-chip coupling losses of the non-optimized grating couplers [48] of approximately 10.5 dB for both interfaces. The on-chip loss of the device is approximately 6 dB for maximum transmission of the modulator and is dominated by scattering losses due to sidewall roughness in the slot waveguides. A more detailed breakdown of the on-chip insertion losses can be found in the Supplementary Information. A reduction of sidewall roughness by optimization of the fabrication process and shorter device lengths is expected to enable SOH MZM with on-chip insertion losses below 2 dB.

Organic material with increased electro-optic activity

In general, the EO coefficient of the organic cladding depends on the nonlinear polarizability of a single chromophore molecule, on the volume concentration of the chromophores in the material, and on the average orientation of the molecules with respect to the applied electric field [135]. This leads to an expression [136] of the form

$$r_{33} = -2\beta_{zzz} N \left\langle \cos^3 \theta \right\rangle \frac{g}{n^4}, \tag{5.3}$$

where β_{zzz} denotes the first-order hyperpolarizability of the molecule along its dipole axis and N is the chromophore volume number density. For a single chromophore molecule, the quantity θ denotes the angle between the molecular dipole axis z and the direction of the external electric field, and the average degree of chromophore orientation is described by the ensemble average $\left\langle \cos^3 \theta \right\rangle$ the so-called acentric order parameter. The quantity g denotes the Lorentz-Onsager local field factor [97] that accounts for partial field screening in the material, and n is the refractive index. Based on Eq. (5.3), the EO coefficient can be enhanced in three ways: First, by increasing the chromophore hyperpolarizability β_{zzz}, second, by maximizing the chromophore density N, and third, by inducing the highest possible acentric orientation $\left\langle \cos^3 \theta \right\rangle$.

Chromophore hyperpolarizability β_{zzz} has continuously been increased over the last years, driven by the advancement of computational methods that allow to theoretically predict the hyperpolarizability of molecular structures [97], [136], [137]. However, chromophore density N and average acentric orientation $\langle \cos^3 \theta \rangle$ are coupled quantities that cannot be maximized independently [97]: Increasing chromophore density leads to strong electrostatic interactions between the dipoles, thereby counteracting the desired acentric orientation of the ensemble. For high densities, these intermolecular interactions lead to partial crystallization of the material, to vanishing EO activity, and to increased scattering loss. The conventional approach to mitigate these interactions is based on using a small chromophore load of typically less than 25 wt.% in a polymer host matrix. SOH integration was so far largely based on these guest-host systems [31], [36], [70], [77], [81], [138], [139], for which in-device EO-coefficients were far below values achieved for parallel-plate poling of bulk reference samples. The incomplete transfer of EO coefficients from bulk materials to devices appears to be a general problem, which does not only occur in SOH devices: EO coefficients of up to 450 pm/V can be achieved in bulk materials [97], whereas the highest reported in-device EO coefficients amount to $r_{33} = 59$ pm/V for SOH devices [99] and to $r_{33} = 138$ pm/V for an all-polymer MZM [100].

In our experiments, the constraints of low in-device EO coefficients are overcome by using the novel monolithic organic material DLD164. The design of the molecule allows to simultaneously increase the chromophore density N and acentric orientation $\langle \cos^3 \theta \rangle$. The chemical structure of DLD164 is depicted in Fig. 5.30(d) and its synthesis is reported in Ref. [69]. The chromophore molecules are designed to have an EO active core (marked in red) with pendant coumarin-containing side groups (marked in blue). The interaction of these side groups effectively suppresses head-to-tail orientation of the dipolar chromophores and thus inhibits partial crystallization of the material. For a more detailed discussion on this principle, the reader is referred to Refs. [28], [69], [140]. This approach allows for applying the neat chromophore as cladding material without the need for an insulating polymer matrix. Moreover, as discussed in Ref. [140], the intermolecular interaction of the coumarin-pendant side groups reduces the number of rotational degrees of freedom of the chromophores from three to two. Based on a theoretical model, this interaction increases the average acentric order $\langle \cos^3 \theta \rangle$ by a factor of two, due to a lower poling energy required to achieve a given average chromophore orientation, see Ref. [28] for a more detailed discussion of this effect.

RESULTS AND DISCUSSION

Poling of the organic material and DC characterization

SOI waveguides are fabricated by standard CMOS processes using optical lithography, and EO materials are deposited by spin coating, see Supplementary Information for a more detailed description. The EO coefficient of the cladding after poling is studied by measuring the voltage-dependent transmission of an MZM at DC, see Fig. 5.31(a). The depicted trace refers to a 1 mm-long device which was poled by a voltage of 40 V per phase modulator section. This corresponds to a remarkably high poling field of $E_{\text{poling}} \approx 250$ V/µm in the 160 nm-wide slots. The ability to withstand extraordinarily high poling fields appears to be a beneficial result of the SOH slot waveguide structure - for a bulk reference sample of monolithic organic cladding material, dielectric breakdown was already observed [69] at 75 V/µm. We attribute the improved stability to thin-film effects and to a low defect probability in the slot region. The devices are characterized at infrared telecommunication wavelengths around 1550 nm. The 1 mm-long device reveals a π-voltage of $U_\pi = 0.5$ V measured at DC under a 4 V bias. The bias avoids partial field screening by free charges in the cladding that may occur at small DC fields [141], see Fig. **5.31**(a). The corresponding voltage-length product amounts to $U_\pi L = 0.5$ Vmm at an optical insertion loss of 6 dB. This is a very promising result, especially when comparing the SOH MZM to a typical carrier-depletion all-silicon modulator [142] that features a π–voltage of $U_\pi = 3.1$ V at a length of 6 mm and an insertion loss of 9 dB. Taking into account the measured waveguide dimensions and the associated overlap between optical mode and RF field, we can derive the in-device EO coefficient of the cladding to be $r_{33} = (180{\pm}20)$ pm/V, see Supplementary Information for details of the calculation. This coefficient clearly exceeds the maximum value of $r_{33} = 137$ pm/V previously achieved in the corresponding bulk material reference [69], where poling fields were limited by dielectric breakdown.

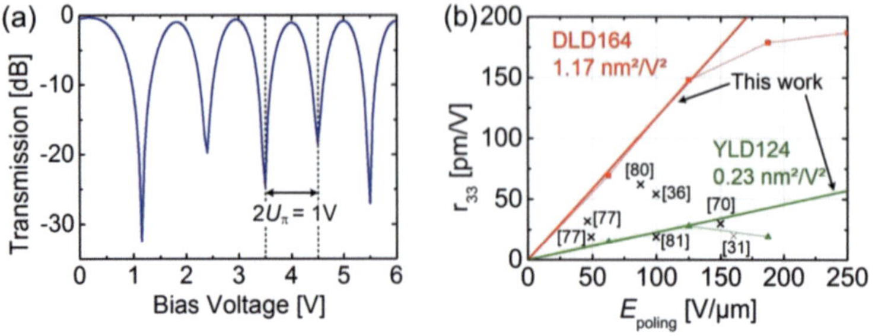

Fig. 5.31 Static characterization of SOH MZM. (a) Transmission vs. applied DC voltage for a 1 mm long SOH MZM in push-pull configuration. At a bias of 4 V, we find a π-voltage of $U_\pi = 0.5$ V. For small voltages, free charges in the cladding lead to partial field screening of the applied fields, which leads to increased π-voltages. (b) EO coefficient r_{33} vs. poling field E_{poling} for a monolithic chromophore cladding (DLD164, red) and for a guest-host cladding containing 25 wt.% YLD124 in a PMMA matrix (green). For weak poling fields, r_{33} increases linearly with the poling field, as indicated by fitted straight lines. DLD164 reveals a poling efficiency of $r_{33}/E_{poling} = 1.17\ \mathrm{nm^2/V^2}$ which is significantly higher than the value of $0.23\ \mathrm{nm^2/V^2}$ obtained for the YLD124/PMMA guest-host cladding. Crosses mark previously reported values for chromophore-doped polymers. For the YLD124/PMMA guest-host cladding, breakdown occurs at field strengths exceeding 200 V/µm, whereas DLD164 withstands all applied voltages, leading to an EO coefficient r_{33} of up to 180 pm/V.

To verify that the extraordinarily high in-device EO coefficient is in agreement with theoretical predictions, we compare the poling efficiency r_{33}/E_{poling} of the DLD164 cladding with that of the guest-host cladding YLD124 (25 wt.%) in PMMA using nominally identical SOI waveguide structures. In terms of molecular hyperpolarizability β_{zzz}, the chromophore cores of DLD164 and YLD124 are identical. However, by utilizing the neat chromophore DLD164, we achieve a 2.3 times higher number density N of EO active dipoles as compared to the commonly used mixture of 25%wt. YLD124 in PMMA, see Supplementary Information for details. For comparison of the acentric order parameters in both materials, we derive the EO coefficients r_{33} for various poling fields, Fig. 5.31(b). For weak poling fields, the EO coefficient increases linearly with E_{poling}. From the measurements, we deduce a poling efficiency of $r_{33}/E_{poling} = 1.17\ (\mathrm{nm/V})^2$ for the DLD164 material and of $0.23\ (\mathrm{nm/V})^2$ for the YLD124/PMMA composite. Using Eq. 1, we find that the product $N\langle\cos^3\theta\rangle$ for DLD164 is increased by a factor of 4.8 as compared to YLD124/PMMA: The calculated quantities β_{zzz} are equal in both cases, while the g/n^4-ratio for

YLD124/PMMA is larger by a factor of 1.07 compared to DLD164 [69]. Considering the 2.3-fold higher chromophore concentration N, we conclude that the acentric order parameter $\langle \cos^3 \theta \rangle$ for DLD164 must be a factor 2.1 times larger than that of the YLD124/PMMA composite, which is in excellent agreement with theoretical predictions [28]. These findings are also consistent with independent experimental investigations of an identical chromophore concept in parallel-plate poled reference measurements [140], revealing enhancements of the acentric order parameter $\langle \cos^3 \theta \rangle$ by a factor 2.1 when using coumarin-based pendant side groups. We hence conclude that the concept of reduced dimensionality through harnessing engineered intermolecular interactions can be directly transferred from parallel-plate poled reference samples to devices and allows to significantly enhancing the performance of SOH devices.

Data transmission and energy consumption

The fabricated SOH modulators were tested in transmission experiments using a non-return-to-zero (NRZ) on-off keying (OOK) format. We perform two sets of experiments, aiming at modulation with lowest possible energy consumption, or with highest possible data rate. Both experiments were performed at a wavelength of 1546 nm. For the low-energy experiments we use a 1 mm-long device operated at data rates of 12.5 Gbit/s and 25 Gbit/s; for the high-speed experiments, we use also devices as short as 250 µm. The corresponding experimental setup is sketched in Fig. 5.32 and is explained in more detail in the Supplementary Information along with the models that are used for estimating the dissipated energy per bit.

For the low-energy experiments, we operate the MZM in different driving modes and measure the bit error ratio (BER) along with the energy consumption per bit as a function of drive voltage, see Fig. 5.33(a) and (b). For high data rates beyond 20 Gbit/s, the 1 mm-long device acts as a travelling-wave modulator and the transmission line needs to be terminated by a matched load resistor of $R_L = 50\,\Omega$ to avoid detrimental back-reflections of the RF wave, Fig. 5.33(c). In this case, the RF power is dissipated along the lossy transmission line and in the termination. For a mean-free drive signal switching between $-U_{drive}/2$ and $+U_{drive}/2$, the energy per bit W_{bit} is obtained by dividing the dissipated average power by the data rate r, $W_{bit} = (U_{drive}/2)^2/(R_L \cdot r)$.

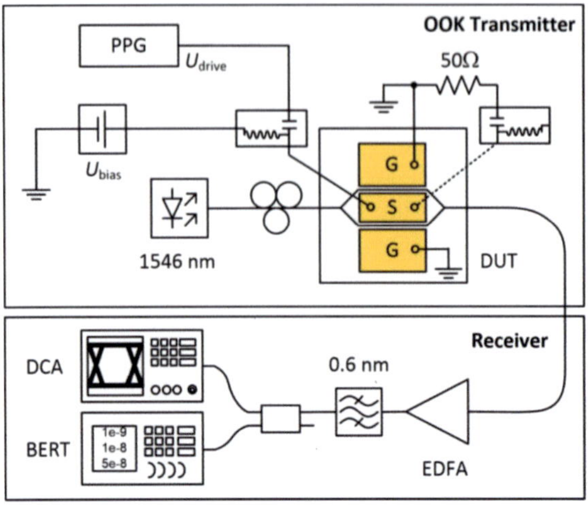

Fig. 5.32 Experimental setup for transmission experiments. A pseudo-random pattern generator (PPG) is used to drive the modulator without additional RF amplifier. The drive signal is fed to the device under test (DUT) by using a ground-signal-ground (GSG) RF probe. A bias-T is used to adjust the operating point. A second RF probe at the end of the modulator can be used to terminate the device. Light from an external-cavity laser is coupled to the MZM using grating couplers. The modulated light is amplified by an erbium-doped fibre amplifier (EDFA) and detected by a digital communications analyzer (DCA) and a bit-error-ratio tester (BERT).

For data rates below 20 Gbit/s, the modulator is much shorter than the RF wavelengths, and the device acts as a "lumped" element. In this case, the MZM may be operated without termination, Fig. 5.33(d). This significantly decreases the energy consumption, since the drive signal is no longer dissipated in a 50 Ω-termination [143]. Instead, energy consumption is now dominated by ohmic losses that occur in the internal resistor R_s of the source and in the series resistor R_d of the device during charging and de-charging of the capacitor [125] C_d, see Fig. 5.33(d) for an equivalent-circuit model. For a terminated device, the drive voltage U_{drive} is half the open-circuit voltage U_0 of the 50 Ω-source, whereas for the non-terminated device, the drive voltage U_{drive} can approach the full open-circuit voltage, see equivalent circuit in Fig. 5.33(d). This effect must be taken into account when calculating the drive voltage and the associated energy consumption of the terminated and non-terminated device. Moreover, the device capacitance C_d is an important parameter for estimating the energy consumption of the unterminated modulator. The device capacitance can be measured using a vector network analyzer (VNA), leading, e.g., to a value of $C_d = (400 \pm 20)$ fF for a 1 mm-long device, see Supplementary Information for a more detailed description of the measurement. The measured results are in good agreement with the values predicted by electrostatic simulations (CST Microwave Studio). In general, the device capacitance C_d is a frequency-dependent quantity due to its dispersive dielectric. When calculating the energy consumption, this can be taken into account by integration over the dissipated power spectrum in the frequency domain, see Supplementary Information.

For small data rates far below the 3 dB cutoff frequency of the modulator, power dissipation associated with charging/de-charging the capacitor C_d from

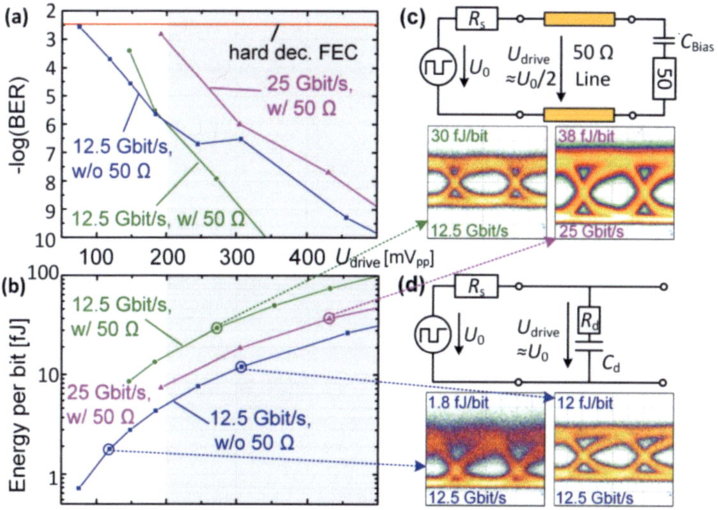

Fig. 5.33 Modulation at low drive voltage. (a) Measured bit error ratio (BER) as a function of drive voltage. For transmission at 25 Gbit/s (magenta) the DUT is always terminated by a 50 Ω load, whereas for a rate of 12.5 Gbit/s, the device is driven with (green) and without (blue) termination. Data points in the gray-shaded areas were measured at the quadrature operating point of the device. (b) Corresponding energy per bit for the various drive voltages of (a). (c) Equivalent circuit of the terminated travelling-wave modulator along with the corresponding eye diagrams measured at the quadrature operating point at 25 Gbit/s ($U_{drive} = 430$ mV$_{pp}$, $W_{bit} = 38$ fJ) and at 12.5 Gbit/s ($U_{drive} = 270$ mV$_{pp}$, $W_{bit} = 30$ fJ). The internal resistance of the source and the terminating impedance amount to $R_s = R_L = 50\ \Omega$. (d) Eye diagrams at 12.5 Gbit/s of a non-terminated device along with the respective equivalent lumped-element circuit. The right inset shows an eye diagram at $U_{drive} = 310$ mV$_{pp}$ ($W_{bit} = 12$ fJ) when operating the MZM at the quadrature point. The left inset shows an eye diagram at $U_{drive} = 120$ mV$_{pp}$ ($W_{bit} = 1.8$ fJ) when operating below the quadrature point.

an initial voltage 0 V to a final voltage $U_{\mathrm{drive}} = U_0$ and vice versa is independent of the series resistance and amounts to [125] $C_{\mathrm{d}}U_0^2/2$. When driving a modulator with a digital NRZ signal, we need to take into account that charging and de-charging only occurs during 1-0 and 0-1 transitions, but not for 0-0 and 1-1 sequences. On average, charging or de-charging of the capacitor occurs every second bit, and the energy consumption per bit hence amounts to [125] $W_{\mathrm{bit}} = C_{\mathrm{d}}U_0^2/4$. For data rates r approaching the 3 dB cut-off frequency of the modulator, however, the device capacitor is no longer fully charged within the timeslot of one symbol, and the energy consumption can no longer be calculated by the simple expression $W_{\mathrm{bit}} = C_{\mathrm{d}}U_0^2/4$. Instead, the dissipated power has to be computed for each frequency component of the NRZ power spectrum separately. The total dissipated power is then obtained by integrating over the corresponding power spectrum, see Supplementary Information for a more detailed description. For a frequency-independent device capacitance C_{d}, this leads to an expression of the form

$$W_{\mathrm{bit}} = \frac{1}{4} C_{\mathrm{d}} U_0^{\,2} \left[1 - e^{-\frac{T}{\tau}} \right] \tag{5.4}$$

In this expression, $T = 1/r$ denotes the symbol duration, $\tau = (R_{\mathrm{s}} + R_{\mathrm{d}})C_{\mathrm{d}}$ is the RC time constant of the circuit, and R_{s} and R_{d} are the series resistances of the source and of the device, as defined in Fig. 5.33(d). This model can be extended to the case of a frequency-dependent capacitance C_{d}, see Supplementary Information. In addition to the low-frequency approximation and the exact frequency-domain model, we also investigate a slightly simplified method based on a time-domain approximation. We find good agreement between the various methods for estimating the energy consumption of our devices and hence conclude that the results are reliable, see Supplementary Information.

For transmission at 12.5 Gbit/s, we measure comparable BER for the terminated and non-terminated device at small drive voltages, see Fig. 5.33(a) and Fig. 5.33(b). The associated energy consumption, however, is significantly lower for the non-terminated device. A drive voltage of only 80 $\mathrm{mV_{pp}}$ is required to keep the measured BER below the hard-decision FEC threshold [144] of 4.5×10^{-3}, corresponding to an energy consumption of 0.7 fJ/bit. For remaining at a BER below 10^{-9}, we find a drive voltage of $U_{\mathrm{drive}} = 460\ \mathrm{mV_{pp}}$ (27 fJ/bit) for the non-terminated MZM and $U_{\mathrm{drive}} = 300\ \mathrm{mV_{pp}}$ (40 fJ/bit) for the terminated MZM. These energy consumptions are one to two orders of magnitude below the 200 fJ/bit which were previously demonstrated for a

5 mm-long silicon-based MZM [23] operated at 630 mV_{pp}. The performance of our MZM is even comparable to best-in-class resonant structures [17], [24], where 3 fJ/bit were reported at drive voltages of 1 V_{pp}. For transmission at 25 Gbit/s, slightly worse BER are measured, and the energy consumption amounts to 7 fJ/bit (190 mV_{pp}) at a BER of 4.5×10^{-3} and to 52 fJ/bit for a BER of 10^{-9}. These figures do not include the energy consumption of any external components, such as laser, optical amplifier and electrical drivers. Currents flowing due to DC voltages are measured to be below 2 nA, contributing only a negligible amount of energy (< 1 aJ/bit) to the budget.

For the high-speed transmission experiments at data rates of up to 40 Gbit/s, we use 500 µm-long and 1 mm-long terminated devices and a 250 µm-long non-terminated device. For the 1 mm-long device, optimum signal quality at 12.5 Gbit/s was obtained for a peak-to-peak drive voltage swing of $U_{\text{drive}} = 950$ mV_{pp} measured at the output of the pulse-pattern generator. This voltage was also used for the other data rates to facilitate comparison. Note

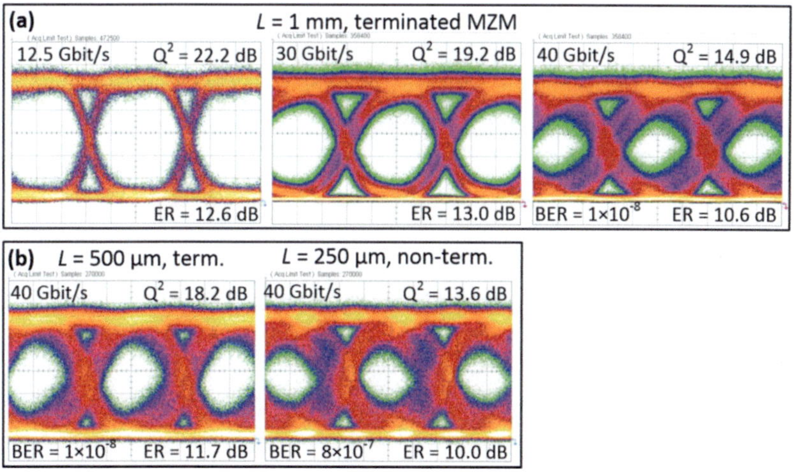

Fig. 5.34 High-speed modulation. (a) NRZ OOK eye diagrams of a 1 mm long modulator for data rates of 12.5 Gbit/s, 30 Gbit/s, and 40 Gbit/s. The modulator is biased at the quadrature point. The drive voltage is 950 mV_{pp}. The extinction ratios (ER) exceed 10 dB even at 40 Gbit/s. Up to 35 Gbit/s the BER is below 1×10^{-12}. At 40 Gbit/s we measure a low BER of 1×10^{-8}. (b) 40 Gbit/s eye diagrams of a 500 µm long terminated modulator ($U_{\text{drive}} = 2$ V_{pp}) and of a 250 µm long modulator without termination ($U_{\text{drive}} = 3.4$ V_{pp}). Measured Q^2 factors, ER and BER are denoted in the respective figures. The PRBS length is 2^{31}-1.

that, due to slight over-modulation and due to the frequency response of the device and the deployed RF components, the drive voltage swing is bigger than 500 mV_{pp} that would be expected from the DC voltage-length product of $U_\pi L = 0.5$ Vmm. Eye diagrams for various data rates are depicted in Fig. 5.34(a). The extinction ratios (ER) exceed 10 dB in all cases. At 12.5 Gbit/s and 30 Gbit/s we measure excellent Q^2-factors of 22 dB and 19 dB, respectively, and obtain error-free operation with BER values below 10^{-12}. At 40 Gbit/s we measure a Q^2 of 15 dB and a reasonably low BER of 1×10^{-8}. We demonstrate that 40 Gbit/s data rates can also be generated with 500 µm-long terminated MZM and with 250 µm-long non-terminated MZM, which were operated with drive voltages of $U_{drive} = 2\ V_{pp}$ and $U_{drive} = 3.4\ V_{pp}$, respectively. These findings demonstrate that SOH modulators support data rates comparable to those of carrier-depletion p-n modulators while featuring small device footprints comparable to those of carrier-injection p-i-n modulators.

CONCLUSIONS

We have shown that SOH integration offers the potential to implement EO devices with unprecedented performance while maintaining many of the advantages of standardized large-scale CMOS processing. We experimentally demonstrate non-resonant EO modulation at 12.5 Gbit/s with a peak-to-peak voltage swing of 80 mV and an energy consumption as small as 0.7 fJ/bit. The in-device EO coefficient of the organic cladding amounts to $r_{33} = (180\pm20)$ pm/V and is significantly larger than the values that can be achieved by parallel-plate poling of the corresponding bulk material. Moreover, our SOH modulators feature small footprint and are capable of operating at 40 Gbit/s. Our findings demonstrate that SOH integration allows to overcome intrinsic limitations of all-silicon devices: The concept combines low $U_\pi L$-products comparable to those of carrier-injection devices with large electro-optic bandwidths known from carrier-depletion devices, while the energy consumption is less by at least one order of magnitude. We believe that there is still significant room for enhancing the performance of SOH modulators, e. g., by taking advantage of the continuously improving EO cladding materials.

One important aspect of organic EO materials is the long-term stability, both regarding thermal and photo-induced degradation. Within the experiments of this paper, we have not encountered any photo-degradation or bleaching effects, even when operating the device at input powers of more than 15 dBm, which corresponds to an intensity of 3×10^6 W/cm^2 in the slot of the SOH modulators. At higher power levels, thermal effects were already beginning to impair the performance of the device. We hence believe that material stability in our device is governed exclusively by thermal reorientation of chromophores. The glass transition temperature of the currently used DLD164 amounts to $T_g = 66°C$, and thermally activated re-orientation of the chromophores reduces the lifetime of the current devices to roughly a month under ambient laboratory conditions. However, we have recently demonstrated that electro-optic coefficients of up to 230 pm/V can also be achieved by other material systems such as binary-chromophore organic glasses that combine shape-engineered dendritic molecular structures with conventional bare donor-acceptor species [141]. These materials feature glass transition temperatures of more than 100 °C and are expected to exhibit better thermal stability. In addition, we expect that stability can be significantly improved by synthetically modified DLD164 chromophores that bear specific crosslinking agents for post-poling lattice hardening. The viability of this approach has already been demonstrated for similar classes [145], [146] of EO compounds, with materials being stable for temperatures of up to 250 °C. It has been demonstrated experimentally that these techniques can be applied to realize temperature-stable all-polymeric MZM [100]. Whereas earlier demonstrations of crosslinking have led to increased optical losses by absorption or scattering in the organic material, it has recently been demonstrated that this problem can be overcome by utilizing cycloaddition reactions (Diels-Alder and related cycloadditions) that do not produce chemical side products or lead to lattice distortion. In these experiments, no increase in loss has been observed in the near infrared [147], and the values obtained for the glass transition temperatures after crosslinking are above 200°C [148]. For DLD164, the addition of crosslinking agents to the side groups can be realized without affecting the EO activity of the chromophore or the coumarin-coumarin interaction of the side chains. Thermal stability and photo-degradation of SOH devices is subject to ongoing research.

Supplementary Information

A. Design and fabrication of silicon-on-insulator (SOI) waveguides

The waveguide structures in this work were fabricated by standard CMOS fabrication processes and 193 nm deep-UV lithography on a 200 mm SOI wafer with 220 nm-thick device layer and 2 µm-thick buried oxide. Multimode interference (MMI) couplers are used as power dividers and combiners for the Mach-Zehnder interferometers. Low-loss logarithmically tapered strip-to-slot mode converters [109] are used to couple the access strip waveguides to the slot waveguides of the phase modulators. The measured slot width is 160 nm and the rail width is 210 nm. The rails are connected to 60 nm-thick n-doped (As, $n_D = 3 \times 10^{17}$ cm^{-1}) silicon slabs. The silicon structures are covered with a 1.1 µm-thick SiO_2 layer and are connected to the 500 nm-thick copper transmission line by 900 nm-high tungsten vias. After electrode fabrication, the oxide was locally opened by dry and wet etching to enable deposition of the organic EO material in the slot region. The optical on-chip loss of the MZM amounts to approximately 6 dB for maximum transmission of the 1 mm long modulator. The loss of the MZM can be decomposed into losses of the passive components, scattering loss due to sidewall roughness in the slot region of the modulator, material absorption loss of the EO organic cladding, and free-carrier absorption in the doped silicon waveguide structures. Passive optical components, such as splitters, combiners, and strip-to-slot mode converters [109] contribute only 0.5 dB to the MZM loss. The material absorption of the bulk EO organic cladding DLD164 amounts to 1.3 dB/mm at infrared telecommunication wavelengths around 1550 nm. Since roughly 50% of the mode field interacts with the organic cladding, material absorption contributes 0.65 dB/mm to the total propagation loss in the phase shifters. For free-carrier absorption in the doped silicon waveguide core [149] ($n_D = 3 \times 10^{17}$ cm^{-3}), we estimate a propagation loss contribution of 0.55 dB/mm, taking again into account that less than 50% of the guided light actually interacts with the doped silicon region. Scattering loss due to rough waveguide sidewalls hence remains as a dominant loss mechanism. The estimated contribution of scattering loss amounts to 4.3 dB/mm, which is mainly caused by fabrication imperfections that occur during etching of the waveguides and during opening of the back-end oxide. We expect that scattering losses can be significantly reduced by optimization of the process parameters. This can, e.g., be achieved by using immersion lithography on 300 mm wafers, which enables slot waveguides with propagation losses of less than 1 dB/mm, see ref. [4]. Similarly, for 200 mm technology, propagation losses below 0.7 dB/mm have been demonstrated [150] by exploiting a larger slot width of 190 nm. Further reduction is possible by deploying asymmetric slot geometries, where losses of 0.2 dB/mm have been demonstrated [151]. Using these techniques and reducing the device length, on-chip modulator losses between 1 dB and 2 dB should be feasible in future device generations.

B. Electro-optic cladding

The nonlinear chromophore DLD164 is applied to the chip by spin coating from a solution of 1,1,2-trichloroethane. Before spin-coating, the solution is exposed to ultrasonic agitation and is subsequently filtered through a membrane filter. The chip is baked under vacuum prior to poling. The material is poled in push-pull orientation at glass-transition temperature by applying a DC voltage across the floating ground electrodes of the transmission line as described in Fig. 5.30(b) and Ref [134]. The refractive index of DLD164 has been measured by ellipsometry [69] on various material samples.

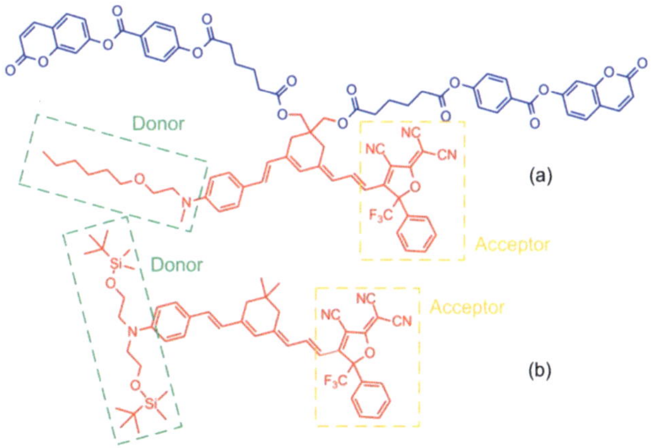

Fig. 5.35 Chemical structure of the electro-optic chromophores DLD164 (a) and YLD124 (b), respectively. The electro-optic active parts are highlighted in red color. The molecular structure of the active segments of DLD164 and YLD124 is identical except for differences in the solubilizing head group on the donor part of each molecule. Since these groups are decoupled from the conjugated system, they do not have any substantial influence on the hyperpolarizability β_{zzz} of the molecule and can thus be neglected.

At a wavelength of 1546 nm, the refractive index ranged from $n = 1.78$ to $n = 1.83$, depending on the sample.

The molecular weight of the DLD164 chromophore amounts to M_w(DLD164) = 1523 g/mol. According to the colour scheme in Fig. 5.35, the electro-optic chromophore core (red) accounts for 705 g/mol (46%), and the side-groups (blue) account for 818 g/mol (54%). Each molecule in the cladding contains one chromophore, thus 1 g of neat DLD164 cladding material contains 6.57×10^{-4} mol chromophores. In the case of YLD124 (molecular weight M_w(YLD124) = 880 g/mol) the final cladding material contains 75 wt.% PMMA, thus 1 g of YLD124/PMMA contains 0.25 g of YLD124, which corresponds to 2.84×10^{-4} mol chromophores. The molar ratio M_w(DLD164)/ M_w(YLD124) is 2.31. Since the mass density of both materials can be considered identical [69], this ratio corresponds as well to the ratio of chromophore number density N in the slot volume.

C. Evaluation of the electro-optic coefficient r_{33}

The EO coefficient of the organic cladding material that fills the slot waveguide can be estimated from the π-voltage U_π of a push-pull MZM, using the expression [47], [130]:

$$r_{33} = \frac{w_{\text{slot}} \lambda_c}{2 L U_\pi \Gamma n_{\text{slot}}^3}, \tag{5.5}$$

where λ_c is the carrier wavelength, w_{slot} is the width of the silicon slot, L is the modulator length, n_{slot} is the refractive index of the EO material in the slot, and Γ is the field interaction factor. The interaction factor Γ is calculated from the electric mode field $\mathcal{E}_x$ of the optical slot-waveguide using the expression [47], [130]

$$\Gamma = \frac{n_{\text{slot}}}{Z_0} \frac{\int_{\text{slot}} \frac{1}{2} |\mathcal{E}_x|^2 \, \mathrm{d} A_{\text{slot}}}{\mathcal{P}(\omega_c)}, \tag{5.6}$$

where $\mathcal{P}$ is the power of the mode field, ω_c the frequency of the optical carrier and Z_0 is the impedance of free space. For the waveguide geometry in this work and the refractive index of DLD164 $n_{\text{slot}} = 1.83$, we calculate a field interaction factor of $\Gamma = 0.21 \pm 0.02$. The uncertainty of Γ is a result of limited accuracy when measuring the device dimensions. The reported values of r_{33} were calculated using the mean value $\Gamma = 0.21$.

D. Data transmission experiment

The experimental setup is depicted in Fig. 5.32. We use laser light at 1546 nm with a typical power of 5 dBm and controlled polarization, which is coupled to the DUT using on-chip grating couplers [48]. Light is modulated by coupling the signal of a pseudo-random bit pattern generator (PPG) to the coplanar transmission line of the modulator using a GSG RF probe. The pattern sequence length amounts to 2^{31}-1. Electrical signals are coupled to the chip by using standard RF probes (GGB Industries, Picoprobe). For high data rates, a second RF probe is used to terminate the transmission line. A bias voltage of 2 V is applied to the GSG electrodes using a bias-T. The modulated light is amplified by an erbium doped fibre amplifier (EDFA) and then received by a digital communications analyzer (DCA) and a bit error ratio tester (BERT). The output power of the EDFA is kept at constant value of approximately 13 dBm throughout the measurements. A gate field of up to 250 V/µm is applied between modulator and substrate for increasing the conductivity of the silicon as explained in Ref. [31].

E. Modulator capacitance measurement

For a non-terminated device, the device capacitance can be extracted from the complex frequency-dependent amplitude reflection factor S_{11} at the device input. The input impedance Z_d is linked to S_{11} by

$$Z_\mathrm{d}(\omega) = Z_\mathrm{L}\frac{1+S_{11}(\omega)}{1-S_{11}(\omega)}\,, \tag{5.7}$$

where $Z_\mathrm{L} = 50\ \Omega$ denotes the impedance of the RF feed line. The frequency-dependence of S_{11} can be measured by using a vector network analyzer (VNA) and performing a reflection factor measurement. A calibration

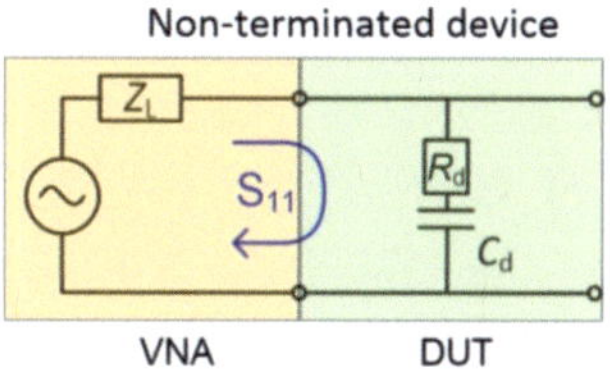

Fig. 5.36 Lumped-element equivalent-circuit model of the VNA measurement, valid for RF wavelengths that are much larger than the modulator length. The reflection factor S_{11} of the non-terminated device is measured. The device impedance Z_d and the device capacitance C_d are extracted from the S_{11} parameter

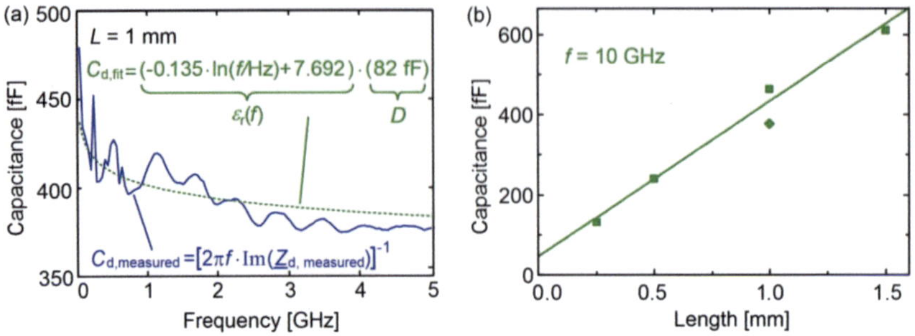

Fig. 5.37 Measurement of device capacitance C_d. (a) Measured capacitance of a 1 mm long MZM (blue) as a function of frequency and fitted function (green dashed curve). (b) Measured device capacitance for different device lengths at 10 GHz extracted from the respective fits. A linear relation between capacitance and length is confirmed.

substrate is used for reference measurements, thereby taking into account all the parasitics of the feed lines and the RF probe.

For low frequencies, the modulator is much shorter than the operating radio-frequency (RF) wavelength and can be treated as a "lumped" element. As a simple model, we assume an equivalent circuit which only comprises a series resistor R_d and a capacitor C_d, as depicted in Fig. 5.36. The device impedance can then be written as

$$Z_d = R_d + \frac{1}{j\omega C_d}. \tag{5.8}$$

For a 1 mm long device, this approximation holds up to approximately 10 GHz, and the device capacitance C_d can directly be extracted from the frequency-dependent imaginary part of the measured device impedance, see blue trace in Fig. 5.37(a). Note that the device capacitance depends on frequency since the silicon slot is filled with a dispersive dielectric. A simple model function, based on the capacitance of a parallel-plate capacitor, $C_{d,fit} = \varepsilon_r(f) \cdot D$, is then fitted to the measured data, where $\varepsilon_r(f)$ is the known permittivity function of the electro-optic organic material and D is the only fitting parameter. The fit is represented by the dashed green trace in Fig. 5.37(a). This way, the capacitance can be extrapolated even for frequencies exceeding 10 GHz.

Capacitance measurements have been repeated for devices of different length, see Fig. 5.37(b). As expected, the relation between capacitance and length is linear. In addition, we confirm the measured capacitance by an electrostatic simulation using the commercial simulation software CST Microwave Studio, leading to a value of 402 fF for a 1 mm long device at a frequency of 40 MHz. This is in good agreement with the corresponding value of 436 fF as obtained from fitting the measurement results. The capacitance has furthermore been measured at a frequency of 1 MHz using an LCR-meter and is found to deviate by only 16 % from the extrapolation of the VNA measurements to low frequencies. We thus conclude that the device capacitance C_d was estimated with reasonable accuracy.

F. Estimation of the energy consumption

Travelling-wave modulator with 50 Ω termination

In the following, we assume signals with ideally rectangular non-return-to-zero (NRZ) pulse shapes, featuring a defined peak-to-peak voltage of U_{drive} at the input of the device. For the travelling-wave modulators, the coplanar transmission line is matched to a 50 Ω source impedance and terminated by a matched load resistor $R_L = 50\ \Omega$, see Fig. 5.33(d) of the main paper for an equivalent-circuit model of the device and of the RF source. The RF power is then dissipated along the lossy transmission line and in the termination, and the energy per bit W_{bit} is obtained by dividing the dissipated power by the data rate r, $W_{bit} = (U_{drive}/2)^2/(R_L \cdot r)$. For the terminated device, the drive voltage U_{drive} is measured directly by replacing the 50 Ω-terminated device by an oscilloscope with 50 Ω input impedance. According to the equivalent circuit in Fig. 5.33(d) the drive voltage is half the value of the open circuit voltage U_0 of the source.

Lumped-element device without termination

The energy consumption of a non-terminated device is dominated by power dissipation in the series resistances R_s and R_d when charging and de-charging the device capacitor C_d, see Fig. 5.33(d) of the main paper for an equivalent-circuit model. The dissipated energy depends on both the drive voltage and the device dynamics. In the following, we derive three different methods for estimating the power dissipation of the non-terminated device: In a simple model (Method 1), the energy consumption of a single switching process is estimated by assuming a step-like drive signal which fully charges the capacitor [125]. This method is only valid for data rates far below the RC cut-off frequency of the device. A more general, but also more complex frequency-domain model (Method 2) allows to estimate the energy consumption also for high data rates and frequency-dependent device parameters. A simple but still reasonably accurate model (Method 3) can be obtained by combining the time-domain analysis according to Method 1 with a frequency-domain transfer function similar to the one used in Method 2. We show that all three methods exhibit comparable results when applied to the devices presented in the main paper, thereby indicating the reliability of the estimated energy consumption. In the limit of low data rates, the methods give identical results.

For our analysis, we represent the device by a lumped-element equivalent circuit according to Fig. 5.38(a). Moreover, unless otherwise noted, we assume bipolar rectangular drive signals, which vary between voltage levels of $-U_0/2$ and $+U_0/2$, Fig. 5.38(b) and (c). It has been shown in Ref. [125] that the drive

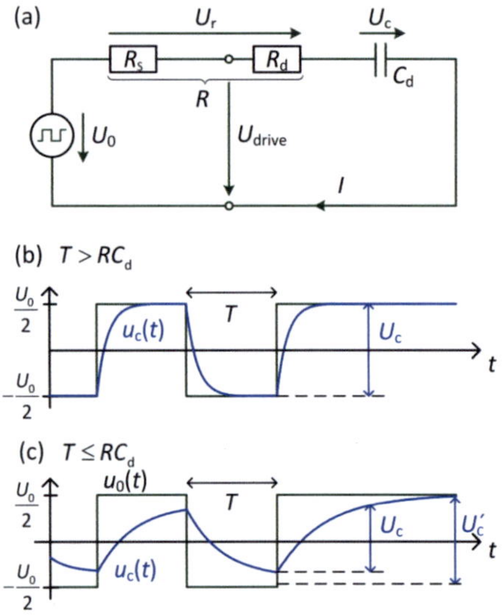

Fig. 5.38 Charging and de-charging of the device capacitor C_d. (a) Equivalent circuit. U_0 denotes the open-circuit voltage of the source, and U_{drive} denotes the drive voltage at the input of the device. The quantities U_c and U_r denote the voltage drops across the device capacitor C_d and the series resistor $R = R_s + R_d$, respectively. (b) Rectangular drive signal (black) and resulting capacitor voltage (blue) for a data rate below the RC cut-off frequency. The device capacitor can be fully charged during the timeslot of one symbol. We assume bipolar rectangular drive signals, which vary between voltage levels of $-U_0/2$ and $+U_0/2$. (c) Rectangular drive signal (black) and resulting capacitor voltage (blue) for a data rate close to the RC cut-off frequency. The device capacitor is no longer fully charged within the timeslot of one symbol. As a consequence, the voltage $U_c < U_0$ across the slot capacitor varies from symbol to symbol.

circuit can be designed to operate the device at any DC bias without additional power dissipation. Hence, the results derived below are also valid for NRZ drive signals, ranging from 0 to U_0, hence having a non-zero DC component of $U_0/2$.

Method 1: *Time-domain analysis for data rates far below the RC cut-off frequency*

For Method 1, we consider a step-like drive signal which fully charges the capacitor from an initial level of $-U_0/2$ to a final level of $+U_0/2$, see Fig. 5.38(b) [125]. The signal source and the device form an RC-circuit, for which the series resistance $R = R_s + R_d$ is given by the sum of the internal source resistance R_s and the device resistance R_d. For a single charging process the energy dissipated in the series resistors ΔW_r amounts to [125]

$$\Delta W_r = \frac{1}{2} C_d U_0^2 \,. \tag{5.9}$$

Note that this expression is independent of the value of the series resistance R. Charging and de-charging occurs statistically every second bit of our pseudo-random bit sequence. Therefore the energy consumption per bit is given by [125]

$$W_{bit} = \Delta W_r / 2 = \frac{1}{4} C_d U_0^2 = \frac{1}{4} C_d U_c^2 \,. \tag{5.10}$$

Here, U_c is the voltage at the capacitor. U_c and U_0 are to be understood as peak-to-peak voltages. Note that Eq. (5.10) is only valid in the limit of low data rates, for which the capacitor is fully charged or de-charged during the time slot of one symbol. It is therefore not clear how accurate the results of this analysis are when applied to our device, which is operated close to its RC cut-off frequency and in addition features a frequency-dependent device capacitance, see Fig. 5.37(a).

Method 2: *General frequency-domain analysis*

To account for limited RF bandwidth and frequency-dependent device parameters, we consider a frequency-domain device model, which is also based on the equivalent circuit depicted in Fig. 5.38(a). According to the Wiener-Khintchine-theorem, the voltage power spectral density $\Theta_0(f)$ of the pseudo-random source signal $u_0(t)$ is given by the Fourier transform of the signal's autocorrelation function $\vartheta_{\mathrm{uu}}(\tau)$,

$$\Theta_0(f) = \int_{-\infty}^{\infty} \vartheta_{\mathrm{uu}}(\tau) e^{-\mathrm{j}2\pi f\tau}\,\mathrm{d}\tau, \quad \vartheta_{\mathrm{uu}}(\tau) = \int_{-\infty}^{\infty} u_0(t) u_0^*(t-\tau)\,\mathrm{d}t. \tag{5.11}$$

Assuming equal probabilities of ones and zeros, the voltage power density $\Theta_0(f)$ of a bipolar rectangular drive signal with peak-to-peak voltage U_0 has the form [152]

$$\Theta_0(f) = \left(\frac{U_0}{2}\right)^2 T \left(\frac{\sin(\pi f T)}{\pi f T}\right)^2, \tag{5.12}$$

where $1/T$ is the symbol rate and where $\Theta_0(f)$ fulfils

$$\int_{-\infty}^{\infty} \Theta_0(f)\,\mathrm{df} = \left(\frac{U_0}{2}\right)^2. \tag{5.13}$$

From the voltage power spectrum $\Theta_0(f)$ at the source, we can calculate the voltage power spectrum $\Theta_{\mathrm{r}}(f)$ at the series resistor R,

$$\Theta_{\mathrm{r}}(f) = |H_{\mathrm{r}}(f)|^2\, \Theta_0(f), \tag{5.14}$$

where $|H_{\mathrm{r}}(f)|^2$ is the respective power transfer function, derived from the voltage divider formula

$$H_{\mathrm{r}}(f) = \frac{1}{1 + 1/\left(\mathrm{j}\,2\pi f R C_{\mathrm{d}}\right)}. \tag{5.15}$$

The physical power that is dissipated in the series resistor R can then be calculated by

$$P_{\mathrm{r}} = \frac{1}{R} \int_{-\infty}^{\infty} \Theta_{\mathrm{r}}(f)\,\mathrm{d}f. \tag{5.16}$$

Note that Eq. (5.16) comprises ohmic power dissipation in both the internal source resistor R_{s} and in the device resistor R_{d}. The average energy consumption per bit W_{bit} is obtained by multiplying the average dissipated power by the symbol duration T,

$$W_{\mathrm{bit}} = P_{\mathrm{r}}T = \frac{T}{R}\int_{-\infty}^{\infty} \left|H(f)\right|^2 \Theta_0(f)\,\mathrm{d}f$$
$$= \frac{T^2}{R}\left(\frac{U_0}{2}\right)^2 \int_{-\infty}^{\infty} \frac{1}{1+1/\left(4\pi^2 f^2 R^2 C_{\mathrm{d}}^{\ 2}\right)} \left(\frac{\sin(\pi f T)}{\pi f T}\right)^2 \mathrm{d}f. \tag{5.17}$$

This expression represents the energy consumption of a non-terminated modulator for arbitrary data rates. Note that the device capacitance C_{d} may be a frequency-dependent quantity, as explicitly shown for the SOH modulator in Fig. 5.37(a). This is directly taken into account in Eq. (5.17) when substituting C_{d} by a frequency-dependent capacitance $C_{\mathrm{d}}(f)$.

When assuming a frequency-independent device capacitance C_{d} and symbol rates $1/T$ far below the RC cut-off frequency, we can show that Eq. (5.17) is equivalent to the result of the simplified time-domain model, Eq. (5.10). With $\tau = RC_{\mathrm{d}}$, Eq. (5.17) takes the form

$$W_{\mathrm{bit}} = \frac{T^2}{2\pi R}\left(\frac{U_0}{2}\right)^2 \int_{-\infty}^{\infty} \frac{1}{1+1/\left(\omega^2\tau^2\right)} \left(\frac{\sin(\omega T/2)}{\omega T/2}\right)^2 \mathrm{d}\omega$$
$$= \frac{T}{\pi R}\left(\frac{U_0}{2}\right)^2 \underbrace{\int_{-\infty}^{\infty} \frac{\sin^2(\omega T/2)}{\left(\omega T/2\right)^2 + \left(T/(2\tau)\right)^2} \mathrm{d}\left(\omega T/2\right)}$$
$$= \frac{T}{\pi R}\left(\frac{U_0}{2}\right)^2 \cdot \left[\frac{2\tau}{T}\pi e^{-\frac{T}{2\tau}} \sinh\left(\frac{T}{2\tau}\right)\right] \tag{5.18}$$
$$= \frac{\tau}{R}\left(\frac{U_0}{2}\right)^2 \left[1 - e^{-\frac{T}{\tau}}\right].$$

Substituting $\tau = RC_{\mathrm{d}}$ leads to the compact expression

$$W_{\mathrm{bit}} = \frac{1}{4} C_{\mathrm{d}} U_0^{\ 2} \left[1 - e^{\frac{-T}{\tau}}\right]. \tag{5.19}$$

Operating the device close to the cut-off frequency hence reduces power dissipation in the series resistor. For $T \gg \tau$, this relation converges to the expected expression of Eq. (5.10)

$$W_{\mathrm{bit}} \overset{T>>\tau}{=} \frac{1}{4} C_{\mathrm{d}} U_0^2 . \tag{5.20}$$

The simplified time-domain model of Method 1 can hence be understood as a spccial case of the general frequency-domain model according to Method 2.

Method 3: *Time-domain approximation for high data rates*

For operation close to the RC cut-off frequency, the capacitor is no longer fully charged or de-charged within the timeslot of one symbol. As a consequence, the capacitor voltage U_c is smaller than the open-circuit source voltage U_0. For estimating the associated power consumption, let us consider the case where the source voltage changes from an initial value of $-U_0/2$ to a value of $U_0/2$ at $t = t_1$, see first symbol in Fig. 5.38(c). In the subsequent symbol slot, the capacitor voltage changes from $-U_c/2$ to $U_c/2$. Following the derivation in Ref. [125], the energy dissipated in the series resistor R during the subsequent symbol slot is then given by

$$\Delta W_{\mathrm{r}} = \int_{t_1}^{t_1+T} u_{\mathrm{r}}(t) i_{\mathrm{r}}(t) \mathrm{d}t = C_{\mathrm{d}} \int_{-U_{\mathrm{c}}/2}^{U_{\mathrm{c}}/2} \left(\frac{U_0}{2} - u_{\mathrm{c}} \right) \mathrm{d}u_{\mathrm{c}} = \frac{1}{2} C_{\mathrm{d}} U_{\mathrm{c}} U_0 . \tag{5.21}$$

The energy consumption of the modulator is then given by

$$W_{\mathrm{bit}} = \frac{1}{4} C_{\mathrm{d}} U_{\mathrm{c}} U_0 . \tag{5.22}$$

This expression also converges to Eq. (5.10) for $T >> \tau$ since $U_{\mathrm{c}} \approx U_0$ in this case.

Thus, the capacitor voltage U_c and the device capacitance C_d determine the energy consumption. However, the capacitor voltage U_c at the end of each symbol slot is varying depending on the sequence of symbols, see Fig. 5.38(c). To estimate the energy consumption, we replace the peak-to-peak capacitor voltage U_c by twice its respective effective voltage $U_{\mathrm{c}} \approx 2U_{\mathrm{c}}^{\mathrm{eff}}$, where the effective capacitor voltage $U_{\mathrm{c}}^{\mathrm{eff}}$ is derived from the corresponding voltage power spectral density at the capacitor $\Theta_{\mathrm{c}}(f)$,

$$U_{\mathrm{c}}^{\mathrm{eff}} = \sqrt{\int_{-\infty}^{\infty} \Theta_{\mathrm{c}}(f) \mathrm{d}f} . \tag{5.23}$$

The quantity $\Theta_{\mathrm{c}}(f)$ can be calculated from the source power spectral density $\Theta_0(f)$ by making use of the voltage divider formula

$$\Theta_{\mathrm{c}}(f) = \left| H_{\mathrm{c}}(f) \right|^2 \Theta_0(f), \tag{5.24}$$

where the power transfer function $\left| H_{\mathrm{c}}(f) \right|^2$ is given by

$$H_{\mathrm{c}}(f) = \frac{1}{1 + \mathrm{j} 2\pi f R C_{\mathrm{d}}} . \tag{5.25}$$

Assuming a frequency-independent capacitance, the energy consumption of the modulator can thus be calculated substituting Eq. (5.23), Eq. (5.24), Eq. (5.25), and the relation $U_c \approx 2U_c^{\mathrm{eff}}$ into Eq. (5.22).

Comparison of the three methods

The three methods for calculating the energy consumption yield similar results when applied to a 1 mm long SOH MZM operated at four different rata rates of 0.5 Gbit/s, 2.5 Gbit/s, 12.5 Gbit/s, and 20 Gbit/s. For calculating the energy consumption according to Method 2, we use the frequency-dependent capacitance of the modulator as depicted in Fig. 5.37(a). The device resistance R_d is extracted from the real part of the measured input impedance and is found to be $R_d \approx 18\Omega$.

Table 6 summarizes the results of the comparison. The second column shows the calculated mean capacitor voltage according to Eq. (5.23). The mean capacitor voltage U_c is approaching the full open circuit voltage U_0 for small data rates, while it can be significantly lower than U_0 for data rates approaching the 3-dB cut-off frequency. The last three columns show the energy consumptions that were estimated by the previously described three methods.

For the calculation according to Method 1, we neglect the frequency dependence of the device capacitance and use its mean value of $C_d = 392$ fF. The energy per bit W_{bit} is thus given by Eq. (5.10) and is independent of the data rate $r = 1/T$. For Method 2, we insert the measured frequency-dependent device capacitance into Eq. (5.17), and we take into account the RC limitations of the device. Method 3 also takes into account the RC limitations of the device, but the frequency dependency of the device capacitance is omitted and a constant value of $C_d = 392$ fF is used for the calculation.

The energy-per-bit values calculated by Method 2 and Method 3 are generally decreasing with increasing data rate r. Method 1, in contrast, overestimates the calculated energy consumption by 10 % to 20 % for data rates close to the cut-off frequency. As expected, for low data rates all three methods provide identical results, since the drive voltage approaches U_0 and the capacitor is fully charged.

Table 6 – Calculated drive voltage and energy consumption of a non-terminated 1 mm long modulator for four different data rates. The modulator resistance is measured to be $R_\mathrm{d} \approx 18\Omega$.

	$U_\mathrm{c} \approx 2U_\mathrm{c}^\mathrm{eff}$	W_bit - Method 1	W_bit - Method 2	W_bit - Method 3
$r = 0.5$ Gbit/s, $U_0 = 400$ mV	397 mV	15.7 fJ	15.7 fJ	15.6 fJ
$r = 2.5$ Gbit/s, $U_0 = 400$ mV	386 mV	15.7 fJ	15.7 fJ	15.1 fJ
$r = 12.5$ Gbit/s, $U_0 = 400$ mV	329 mV	15.7 fJ	14.6 fJ	12.9 fJ
$r = 20$ Gbit/s, $U_0 = 400$ mV	294 mV	15.7 fJ	12.9 fJ	11.5 fJ
$r = 12.5$ Gbit/s, $U_0 = 90$ mV	74 mV	0.8 fJ	0.7 fJ	0.7 fJ

Given this agreement, we thus conclude that the proposed methods for calculating the effective energy consumptions are valid and accurate. The energy consumption values shown in the main text were calculated based on Eq. (5.17) of Method 2, since this method also takes into account the frequency-dependency of the device capacitance and is therefore expected to produce the most accurate results.

Calculation of the effective drive voltage

In our experiment, terminated and non-terminated devices are operated with various different drive voltages U_drive. The drive voltage U_drive is defined as the voltage which drops across both the device resistor R_d and the device capacitor C_d, see Fig. 5.38(a). These drive voltages depend both on the device parameters and on the operating frequency and can be estimated as twice the effective input voltages in analogy to Eq. (5.23),

$$U_\mathrm{drive} \approx 2U_\mathrm{drive}^\mathrm{eff} = 2\sqrt{\int_{-\infty}^{\infty} \Theta_\mathrm{drive}(f)\,\mathrm{d}f}\,. \tag{5.26}$$

The drive voltage power spectral density $\Theta_\mathrm{drive}(f)$ can be calculated from the source spectral density $\Theta_0(f)$

$$\Theta_\mathrm{drive}(f) = \left|H_\mathrm{drive}(f)\right|^2 \Theta_0(f) \tag{5.27}$$

using the transfer function

$$H_\mathrm{drive}(f) = \frac{1 + \mathrm{j}2\pi f R_\mathrm{d} C_\mathrm{d}}{1 + \mathrm{j}2\pi f R C_\mathrm{d}}\,. \tag{5.28}$$

Eq. (5.26) is used to calculate the drive voltages that define the *x*-axis of Fig. 5.33(b) and (c) of the main text.

For the 250 µm long MZM and for a data rate of 40 Gbit/s, we find a mean drive voltage of $U_{\text{drive}} = 0.84\,U_0$, and the same ratio is obtained for the 1 mm long MZM operated at a data rate of 12.5 Gbit/s. In the experiment, the mean drive voltage is determined by measuring half of the open-circuit voltage of the generator with a 50 Ω-terminated scope ($U_{\text{scope}} = U_0/2$) and by multiplying U_0 with the calculated factor of 0.84.

[End of Paper J15]

5.8 Summary

We demonstrated that silicon-organic hybrid Mach-Zehnder modulators are promising building blocks for compact, high-speed, and energy-efficient PICs. The presented modulators have a compact size between 250 µm to 1.5 mm and a small π-voltage-length-product down to 0.5 Vmm. The measured bandwidth of the modulators is 25 GHz, in good agreement with predictions from an analytical model. The MZM is suitable for 40 Gbit/s OOK operation, and works up to 84 Gbit/s with bipolar ASK modulation. At 12.5 Gbit/s the modulators feature an energy consumption of down to 0.7 fJ/bit.

6 Summary and Outlook

In the following the main results and achievements of this work are summarized, and an outlook on future work on energy-efficient photonic integrated circuits is given.

6.1 Summary

This work studies concepts to reduce the energy consumption of silicon photonic integrated circuits. Explicitly this study comprises a novel PIC fabrication scheme based on anisotropic wet etching to reduce propagation loss of nanowire waveguides, the development of an efficient horizontal coupling scheme between a standard singlemode fiber and a surface emitting grating coupler, and the development of a high-speed, low drive voltage silicon-organic hybrid modulator that utilizes the EO properties of a specifically tailored organic material.

Wet Chemical Waveguide Fabrication

A novel silicon waveguide fabrication scheme based on anisotropic wet etching in aqueous potassium hydroxide solutions has been developed. Due to the strong etching anisotropy the resulting sidewall roughness of the waveguides is hardly affected by the initial roughness from the utilized lithography method. The fabricated waveguides feature a trapezoidal cross-section and a sidewall rms roughness of 1 nm to 3 nm. The measured propagation losses of these waveguides are in the order of 6 dB/cm. This measured loss is indeed larger than losses of waveguides that are fabricated in CMOS-foundries, typically in the order of (2…3) dB/cm. However, this scheme does not utilize expensive deep UV lithography systems and dry etching systems. So far, the eBeam system and photoresist have not been optimized for low structure roughness. The scheme can basically be utilized by any academic institution that has access to a suitable lithography system and represents a low-cost solution for the fabrication of silicon PICs. Furthermore a special lithographic design has been demonstrated that enables fabrication of nanowire waveguides by using inexpensive conventional optical lithography [C5]. In addition, the demonstrated scheme enables the fabrication of trapezoidal grating couplers with 4.5 dB coupling loss. Simulations predict that coupling losses to SSMF can be as low as 0.3 dB when using a non-uniform grating design with metallic bottom reflector.

Horizontal Fiber-Chip Coupling

A fiber-chip coupling concept has been demonstrated that utilizes surface emitting grating couplers and enables the in-plane assembly of PIC and optical fiber by making use of special angle-polished fibers. The scheme features a coupling penalty of 0.45 dB when compared to a conventional vertically coupled cleaved fiber. Coupling losses can be below 1 dB when adapting the grating coupler to the radiated beam profile of the angled fiber. The scheme features a high misalignment tolerance. A power penalty of less than 1 dB is expected for a lateral misalignment of ±2.5 μm and for a rotational misalignment of ±2°.

Silicon-Organic Hybrid Modulator

SOH modulators that utilize the optical properties of an EO organic cladding material are well suited for the realization of high-speed optical interconnects with low drive voltage and low switching energies. This has been shown by utilizing a push-pull Mach-Zehnder modulator structure. The passive components of the PIC such as MMI couplers and strip-to-slot mode converters have been carefully designed for low optical insertion loss, with measured losses of 0.1 dB and 0.02 dB [J4, C15], respectively. The bandwidth of the fabricated devices has been analyzed using an analytic model and is measured to be around 25 GHz. This enables high-speed operation of the devices at symbol rates of up to 40 GBd. Using bipolar 8ASK, a data rate of a single MZM of 84 Gbit/s has been demonstrated [J5, C21]. For low power applications we optimized the device geometry [J7] and investigated novel organic material approaches for SOH integration [C28]. This way record-high in-device EO coefficients of up to 230 pm/V [C32] were achieved that enable ultra-low π-voltage-length products of 0.5 Vmm, and a record-low energy consumption of down to 0.6 fJ/bit [C28]. The combination of compact size, low drive voltage and low energy consumption makes the SOH approach superior to all-silicon approaches and particularly interesting for the realization of next generation optical interconnects.

6.2 Outlook

The achievements of this work concerning speed, compact size, and energy-efficiency of SOH modulators help to fulfill the demands of next generation optical interconnects. However, for a practical application the long-term stability of the organic material is of high importance. The glass transition temperature of the investigated materials is still relatively low, typically between 66 °C and 110°C, leading to thermally activated re-orientation of the chromophores and thus to a degradation of the EO coefficient of the material. Future studies need to focus on aging and temperature stability of organic materials. Depending on the application and the required packaging process different critical temperatures are specified. For telecom applications the so-called Telcordia certification demands a stable device performace at a temperature of 85 °C for a duration of 2000 h. Temperature requirements can be even more severe when packaging relies on flip-chip bonding, where soldering temperatures of 240 °C are common.

We expect that material stability can be significantly improved by synthetically modified chromophores that bear specific crosslinking agents for post-poling lattice hardening, or by increasing the molar mass of the chromophores. The viability of the first approach has already been demonstrated for EO compounds [145], [146], where a material stability of up to 250 °C has been achieved.

Due to the large in-device EO coefficients of organic materials and the resulting small drive voltage requirement of SOH modulators that have been found in this thesis, future work on SOH modulators may focus on enhancing the EO bandwidth of the devices in trade of an increase of drive voltage, e.g. by using dual-drive MZM configurations or by using a parallel configuration of the slot capacitors instead of a serial configuration that was used in this work.

Appendix A Mathematical Relations

A.1 Fourier Transformation

The Fourier transformation of a field $\vec{E}$ is given by

$$\tilde{\vec{E}}(\vec{r},\omega)=\mathcal{F}\left\{\vec{E}(\vec{r},t)\right\}=\int_{-\infty}^{\infty}\vec{E}(\vec{r},t)\mathrm{e}^{-\mathrm{j}\omega t}\,\mathrm{d}t \tag{A.1}$$

$$\vec{E}(\vec{r},t)=\mathcal{F}^{-1}\left\{\tilde{\vec{E}}(\vec{r},\omega)\right\}=\frac{1}{2\pi}\int_{-\infty}^{\infty}\tilde{\vec{E}}(\vec{r},\omega)\mathrm{e}^{\mathrm{j}\omega t}\,\mathrm{d}\omega \tag{A.2}$$

A.2 Kramers-Kronig Relations

The susceptibility $\tilde{\chi}^{(1)}(\omega)=\tilde{\chi}_{\mathrm{r}}^{(1)}(\omega)+\mathrm{j}\tilde{\chi}_{\mathrm{i}}^{(1)}(\omega)$ is the Fourier transform of the causal influence function $\chi^{(1)}(t)$. Therefore $\tilde{\chi}^{(1)}(\omega)$ is an analytic spectral function, and its real and imaginary part are linked by a Hilbert transform (Kramers-Kronig relations)

$$\tilde{\chi}_{\mathrm{r}}^{(1)}(\omega)=-\frac{1}{\pi}\mathcal{P}\int_{-\infty}^{\infty}\frac{\tilde{\chi}_{\mathrm{i}}^{(1)}(\Omega)}{\Omega-\omega}\mathrm{d}\Omega, \tag{A.3}$$

$$\tilde{\chi}_{\mathrm{i}}^{(1)}(\omega)=\frac{1}{\pi}\mathcal{P}\int_{-\infty}^{\infty}\frac{\tilde{\chi}_{\mathrm{r}}^{(1)}(\Omega)}{\Omega-\omega}\mathrm{d}\Omega, \tag{A.4}$$

where $\mathcal{P}$ denotes the Cauchy principle value.

Appendix B Mode Coupling

B.1 Maxwell's Equations

Electromagnetic wave propagation in dielectric waveguides is described by Maxwell's equations. In the absence of free charges and currents Maxwell's equations take the form

$$\vec{\nabla} \times \vec{E}(\vec{r},t) = -\frac{\partial \vec{B}(\vec{r},t)}{\partial t} \tag{B.1}$$

$$\vec{\nabla} \times \vec{H}(\vec{r},t) = \frac{\partial \vec{D}(\vec{r},t)}{\partial t}, \tag{B.2}$$

$$\vec{\nabla} \cdot \vec{D}(\vec{r},t) = 0, \tag{B.3}$$

$$\vec{\nabla} \cdot \vec{B}(\vec{r},t) = 0, \tag{B.4}$$

where $\vec{r} = (x, y, z)^{\mathrm{T}}$ is the position vector, and $\vec{E}$ and $\vec{H}$ are the electric and magnetic field vectors. For non-magnetic materials the relation between the magnetic flux $\vec{B}$ and the magnetic field strength $\vec{H}$ is

$$\vec{B}(\vec{r},t) = \mu_0 \vec{H}(\vec{r},t), \tag{B.5}$$

where $\mu_0 = 1.25664 \times 10^{-6}\,\mathrm{Vs/(Am)}$ is the magnetic permeability of vacuum. With the electric polarization vector $\vec{P}$, the displacement $\vec{D}$ is written as

$$\vec{D}(\vec{r},t) = \varepsilon_0 \vec{E}(\vec{r},t) + \vec{P}(\vec{r},t), \tag{B.6}$$

where $\varepsilon_0 = 8.85419 \times 10^{-12}\,\mathrm{As/(Vm)}$ is the electric permittivity of vacuum.

B.2 Guided Modes

For monochromatic light in a lossless, longitudinally invariant, dielectric waveguide, Maxwell's equations are solved by a separation ansatz for the νth mode,

$$\vec{E}_\nu(\vec{r},t) = A_\nu\, \vec{\mathcal{E}}_\nu(x,y,\omega_c) \exp\left[\mathrm{j}\left(\omega_c t - \beta_\nu z\right)\right], \tag{B.7}$$

$$\vec{H}_\nu(\vec{r},t) = \underbrace{A_\nu}_{\text{Ampl.}}\, \underbrace{\vec{\mathcal{H}}_\nu(x,y,\omega_c)}_{\text{transv. mode}} \underbrace{\exp\left[\mathrm{j}\left(\omega_c t - \beta_\nu z\right)\right]}_{\text{propagator}}, \tag{B.8}$$

where z is the direction of light propagation, β_ν is the longitudinal wave number, A_ν is the amplitude of the wave, and $\vec{\mathcal{E}}_\nu$ and $\vec{\mathcal{H}}_\nu$ are the electric and magnetic transverse vectors of the mode field, respectively. The mode fields fulfil the orthogonality relation ([153], Eqs. (8.5-8) and (1.2-12))

$$\frac{1}{4}\iint_{-\infty}^{\infty}\left(\vec{\mathcal{E}}_{\nu}\times\vec{\mathcal{H}}_{\mu}^{*}\right)+\left(\vec{\mathcal{E}}_{\mu}^{*}\times\vec{\mathcal{H}}_{\nu}\right)\cdot\vec{e}_{z}\,\mathrm{d}x\,\mathrm{d}y=\mathcal{P}_{\mu}\delta_{\nu\mu}, \tag{B.9}$$

where μ and ν are the mode numbers, $\mathcal{P}_{\mu}$ is the normalizing modal power (physical cross-section power $\left|A_{\nu}\right|^{2}\mathcal{P}_{\mu}$), and $\delta_{\nu\mu}$ is the Kronecker delta. Substituting Eqs. (B.7), (B.8) in Eqs. (B.1), (B.2) leads to

$$\vec{\nabla}\times\left(\vec{\mathcal{E}}_{\nu}(x,y,\omega_{\mathrm{c}})\exp\left[-\mathrm{j}\beta_{\nu}z\right]\right)=-\mathrm{j}\omega\mu_{0}\vec{\mathcal{H}}_{\nu}(x,y,\omega_{\mathrm{c}})\exp\left[-\mathrm{j}\beta_{\nu}z\right], \tag{B.10}$$

$$\vec{\nabla}\times\left(\vec{\mathcal{H}}_{\nu}(x,y,\omega_{\mathrm{c}})\exp\left[-\mathrm{j}\beta_{\nu}z\right]\right)=\mathrm{j}\omega\varepsilon_{0}\varepsilon_{\mathrm{r}}(x,y)\vec{\mathcal{E}}_{\nu}(x,y,\omega_{\mathrm{c}})\exp\left[-\mathrm{j}\beta_{\nu}z\right]. \tag{B.11}$$

B.3 Poynting Vector and Power Flow

The power flow density of an electromagnetic wave is represented by the Poynting vector $\vec{S}(\vec{r},t)$,

$$\vec{S}(\vec{r},t)=\vec{E}(\vec{r},t)\times\vec{H}(\vec{r},t). \tag{B.12}$$

For monochromatic electromagnetic waves that oscillate with an angular frequency ω_{c}, the time-averaged power flow density vector, also known as the complex Poynting vector $\vec{S}(\vec{r},\omega_{\mathrm{c}})$, is formulated as a cross product of the corresponding complex field amplitudes (Ref. [153], Eq. 1.2-20)

$$\left\langle\vec{S}(\vec{r},t)\right\rangle=\vec{S}(\vec{r},\omega_{\mathrm{c}})=\frac{1}{2}\left(\vec{\tilde{E}}(\vec{r},\omega_{\mathrm{c}})\times\vec{\tilde{H}}^{*}(\vec{r},\omega_{\mathrm{c}})\right). \tag{B.13}$$

The factor $\frac{1}{2}$ is due to time-averaging. The power P which flows through a surface A with an outward-directed normal unit vector $\vec{n}$ at every point is computed by integrating over the real part of $\vec{S}(\vec{r},\omega_{\mathrm{c}})$,

$$\begin{aligned}P&=\int_{A}\mathrm{Re}\left\{\left\langle\vec{S}(\vec{r},t)\right\rangle\right\}\cdot\vec{n}\,\mathrm{d}A\\&=\frac{1}{4}\int_{A}\left(\vec{\tilde{E}}(\vec{r},\omega_{\mathrm{c}})\times\vec{\tilde{H}}^{*}(\vec{r},\omega_{\mathrm{c}})+\vec{\tilde{E}}^{*}(\vec{r},\omega_{\mathrm{c}})\times\vec{\tilde{H}}(\vec{r},\omega_{\mathrm{c}})\right)\vec{n}\,\mathrm{d}A.\end{aligned} \tag{B.14}$$

Next, an abrupt change of the waveguide geometry and is treated.

B.4 Mode Coupling

Many components comprise an abrupt change of the waveguide (WG) geometry. Examples are, e. g., grating couplers [48] for fiber-chip coupling, or multimode interference couplers (MMI) [154] that serve as power dividers. Let us consider an abrupt transition at $z = z'$ from waveguide geometry A to waveguide geometry B along the propagation direction z of the wave. Coupling of the guided input field at WG A to the modes of WG B can be expressed by a superposition of guided modes $\vec{\mathcal{E}}_{\nu,\mathrm{B}}$ and non-guided modes (NG) of WG B,

$$\vec{\tilde{E}}_{\mathrm{A}} = \sum_{\nu} A_{\nu}(z')\vec{\mathcal{E}}_{\nu,\mathrm{B}}(x,y,\omega_{\mathrm{c}})\exp[-\mathrm{j}\beta_{\nu}z] + \mathrm{NG}, \tag{B.15}$$

$$\vec{\tilde{H}}_{\mathrm{A}} = \sum_{\nu} A_{\nu}(z')\vec{\mathcal{H}}_{\nu,\mathrm{B}}(x,y,\omega_{\mathrm{c}})\exp[-\mathrm{j}\beta_{\nu}z] + \mathrm{NG}, \tag{B.16}$$

where the amplitudes $A_{\nu}(z')$ represent the complex coupling coefficients at $z = z'$. In the following we neglect the excitation of non-guided modes. From Eq. (B.15) and Eq. (B.16) the cross-sectional power can be calculated using Eq. (B.14) and the normal unit vector $\vec{e}_z$ directed along the propagation direction z,

$$\begin{aligned}
P &= \frac{1}{4}\int_A \left(\vec{\tilde{E}}_{\mathrm{A}} \times \vec{\tilde{H}}_{\mathrm{A}}^{*} + \vec{\tilde{E}}_{\mathrm{A}}^{*} \times \vec{\tilde{H}}_{\mathrm{A}}\right)\cdot \vec{e}_z \,\mathrm{d}A \\
&= \sum_{\nu}\sum_{\mu} A_{\nu}A_{\mu}^{*}\,\mathrm{e}^{-\mathrm{j}(\beta_{\nu}-\beta_{\mu})z}\,\frac{1}{4}\int_{-\infty}^{+\infty} \vec{\mathcal{E}}_{\nu,\mathrm{B}} \times \vec{\mathcal{H}}_{\mu,\mathrm{B}}^{*}\cdot \vec{e}_z \,\mathrm{d}A \\
&\quad + \sum_{\nu}\sum_{\mu} A_{\nu}^{*}A_{\mu}\,\mathrm{e}^{+\mathrm{j}(\beta_{\nu}-\beta_{\mu})z}\,\frac{1}{4}\int_{-\infty}^{+\infty} \vec{\mathcal{E}}_{\nu,\mathrm{B}}^{*} \times \vec{\mathcal{H}}_{\mu,\mathrm{B}}\cdot \vec{e}_z \,\mathrm{d}A \\
&= \sum_{\nu}\sum_{\mu} A_{\nu}A_{\mu}^{*}\,\mathrm{e}^{-\mathrm{j}(\beta_{\nu}-\beta_{\mu})z}\,\underbrace{\frac{1}{4}\int_{-\infty}^{+\infty}\left(\vec{\mathcal{E}}_{\nu,\mathrm{B}} \times \vec{\mathcal{H}}_{\mu,\mathrm{B}}^{*} + \vec{\mathcal{E}}_{\mu,\mathrm{B}}^{*} \times \vec{\mathcal{H}}_{\nu,\mathrm{B}}\right)\cdot \vec{e}_z \,\mathrm{d}A}_{\mathcal{P}_{\nu}\delta_{\mu\nu}} \\
&= \sum_{\nu} |A_{\nu}|^2 \mathcal{P}_{\nu}.
\end{aligned} \tag{B.17}$$

The terms $\left|A_\nu\right|^2 \mathcal{P}_\nu$ thus express the amount of power that is guided in each mode ν, respectively. An expression for $\left|A_\nu\right|^2$ can be obtained by forming the cross product of $\tilde{\vec{E}}_\mathrm{A}$ of Eq. (B.15) with $\vec{\mathcal{H}}^*_{\nu,\mathrm{B}}$ and by adding the cross product of $\vec{\mathcal{E}}^*_{\nu,\mathrm{B}}$ with $\tilde{\vec{H}}_\mathrm{A}$ of Eq. (B.16). After integrating the sum over the cross-section A and applying then the orthogonality relation Eq. (B.9) we find:

$$\begin{aligned}
&\int_{-\infty}^{+\infty}\left(\tilde{\vec{E}}_\mathrm{A}\times\vec{\mathcal{H}}^*_{\nu,\mathrm{B}}+\vec{\mathcal{E}}^*_{\nu,\mathrm{B}}\times\tilde{\vec{H}}_\mathrm{A}\right)\cdot\vec{e}_z\,\mathrm{d}A \\
&=\sum_\mu A_\mu\,\mathrm{e}^{-\mathrm{j}\beta_\mu z}\int_{-\infty}^{+\infty}\left(\vec{\mathcal{E}}_{\mu,\mathrm{B}}\times\vec{\mathcal{H}}^*_{\nu,\mathrm{B}}+\vec{\mathcal{E}}^*_{\nu,\mathrm{B}}\times\vec{\mathcal{H}}_{\mu,\mathrm{B}}\right)\cdot\vec{e}_z\,\mathrm{d}A \\
&=\sum_\mu A_\mu\,\mathrm{e}^{-\mathrm{j}\beta_\mu z}\,4\delta_{\mu\nu}\mathcal{P}_\mu=4A_\nu\,\mathrm{e}^{-\mathrm{j}\beta_\nu z}\,\mathcal{P}_\nu.
\end{aligned}\tag{B.18}$$

Thus $\left|A_\nu\right|^2$ is given by

$$\left|A_\nu\right|^2=\frac{\left|\frac{1}{4}\int_{-\infty}^{+\infty}\left(\tilde{\vec{E}}_\mathrm{A}\times\vec{\mathcal{H}}^*_{\nu,\mathrm{B}}+\vec{\mathcal{E}}^*_{\nu,\mathrm{B}}\times\tilde{\vec{H}}_\mathrm{A}\right)\cdot\vec{e}_z\,\mathrm{d}A\right|^2}{\mathcal{P}_\nu^2}.\tag{B.19}$$

The integral in the numerator of Eq. (B.19) is named overlap integral. The power coupling coefficient $\kappa_\nu^{(P)}$, which denotes the fraction of input power P that is coupled to mode ν can be obtained from Eqs. (B.17) and (B.19):

$$\begin{aligned}
\kappa_\nu^{(P)}&=\frac{\left|A_\nu\right|^2\mathcal{P}_\nu}{P} \\
&=\frac{\left|\frac{1}{4}\int_{-\infty}^{+\infty}\left(\tilde{\vec{E}}_\mathrm{A}\times\vec{\mathcal{H}}^*_{\nu,\mathrm{B}}+\vec{\mathcal{E}}^*_{\nu,\mathrm{B}}\times\tilde{\vec{H}}_\mathrm{A}\right)\cdot\vec{e}_z\,\mathrm{d}A\right|^2}{P\mathcal{P}_\nu} \\
&=\frac{\left|\int_{-\infty}^{+\infty}\left(\tilde{\vec{E}}_\mathrm{A}\times\vec{\mathcal{H}}^*_{\nu,\mathrm{B}}+\vec{\mathcal{E}}^*_{\nu,\mathrm{B}}\times\tilde{\vec{H}}_\mathrm{A}\right)\cdot\vec{e}_z\,\mathrm{d}A\right|^2}{\int_{-\infty}^{+\infty}\left(\tilde{\vec{E}}_\mathrm{A}\times\tilde{\vec{H}}^*_\mathrm{A}+\tilde{\vec{E}}^*_\mathrm{A}\times\tilde{\vec{H}}_\mathrm{A}\right)\cdot\vec{e}_z\,\mathrm{d}A\int_{-\infty}^{+\infty}\left(\vec{\mathcal{E}}_{\nu,\mathrm{B}}\times\vec{\mathcal{H}}^*_{\nu,\mathrm{B}}+\vec{\mathcal{E}}^*_{\nu,\mathrm{B}}\times\vec{\mathcal{H}}_{\nu,\mathrm{B}}\right)\cdot\vec{e}_z\,\mathrm{d}A}.
\end{aligned}\tag{B.20}$$

Appendix C Perturbed Waveguide

In the following, we will use a general perturbation ansatz to derive an expression for the change of the wave propagation constant $\delta\beta$ in an electro-optic modulator, where the Pockels effect induces a change of the permittivity $\Delta\varepsilon$. The derivation is based on Refs. [39], [155], [156].

We begin with a perturbation ansatz for a perturbed permittivity $\varepsilon_{\mathrm{p}}(x,y,z)$ that has the following form

$$\varepsilon_{\mathrm{p}}(x,y,z) = \varepsilon_{\mathrm{r}}(x,y) + \Delta\varepsilon(x,y,z)\,. \tag{C.1}$$

In the case of a Pockels effect modulator (Section 2.1.3), this change of permittivity is due to the presence of an external modulation field E_{m}. For lumped element modulators the electric modulation field and the resulting change in permittivity $\Delta\varepsilon(x,y)$ can be considered to be longitudinally invariant. For simplicity we assume that the permittivity is a scalar quantity. Substituting in Maxwell's equations (B.1) and (B.2) the perturbed permittivity, using a separation ansatz for the electric and magnetic fields (Eq. (B.7), Eq. (B.8)), and allowing for multiple modes results in

$$\begin{aligned} &\vec{\nabla} \times \sum_{\nu} A_{\nu}(z)\vec{\mathcal{E}}_{\nu}(x,y,\omega_{\mathrm{c}})\exp(-\mathrm{j}\beta_{\nu}z) \\ &= -\mathrm{j}\omega\mu_0 \sum_{\nu} A_{\nu}(z)\vec{\mathcal{H}}_{\nu}(x,y,\omega_{\mathrm{c}})\exp(-\mathrm{j}\beta_{\nu}z)\,, \end{aligned} \tag{C.2}$$

$$\begin{aligned} &\vec{\nabla} \times \sum_{\nu} A_{\nu}(z)\vec{\mathcal{H}}_{\nu}(x,y,\omega_{\mathrm{c}})\exp(-\mathrm{j}\beta_{\nu}z) \\ &= \mathrm{j}\omega\varepsilon_0 \left[\varepsilon_{\mathrm{r}}(x,y) + \Delta\varepsilon(x,y,z)\right] \sum_{\nu} A_{\nu}(z)\vec{\mathcal{E}}_{\nu}(x,y,\omega_{\mathrm{c}})\exp(-\mathrm{j}\beta_{\nu}z). \end{aligned} \tag{C.3}$$

Here, $\vec{\mathcal{E}}_{\nu}$, $\vec{\mathcal{H}}_{\nu}$ are the modal fields of the unperturbed waveguide. Note that the complex field amplitude $A_{\nu}(z)$ now depends explicitly on the propagation direction, so that the perturbed modes are no longer orthogonal and exchange energy. For performing the partial derivatives, we use the identity $\vec{\nabla} \times (\Phi\vec{F}) = \Phi(\vec{\nabla} \times \vec{F}) + (\vec{\nabla}\Phi \times \vec{F})$, where $\Phi = A_{\nu}$ is a scalar and $\vec{F} = \vec{\mathcal{E}}_{\nu}(x,y)\exp(-\mathrm{j}\beta_{\nu}z)$, $\vec{F} = \vec{\mathcal{H}}_{\nu}(x,y)\exp(-\mathrm{j}\beta_{\nu}z)$ are vectors, respectively,

$$\begin{aligned} &\sum_{\nu}\left[A_{\nu}\left(\vec{\nabla} \times \left(\vec{\mathcal{E}}_{\nu}\exp(-\mathrm{j}\beta_{\nu}z)\right)\right) + \left(\frac{\partial A_{\nu}}{\partial z}\vec{e}_z\right) \times \left(\vec{\mathcal{E}}_{\nu}\exp(-\mathrm{j}\beta_{\nu}z)\right)\right] \\ &\qquad = -\mathrm{j}\omega\mu_0 \sum_{\nu} A_{\nu}\vec{\mathcal{H}}_{\nu}\exp(-\mathrm{j}\beta_{\nu}z) \end{aligned} \tag{C.4}$$

$$\sum_\nu\left[A_\nu\left(\vec{\nabla}\times\left(\vec{\mathcal{H}}_\nu\exp(-\mathrm{j}\beta_\nu z)\right)\right)+\left(\frac{\partial A_\nu}{\partial z}\vec{e}_z\right)\times\left(\vec{\mathcal{H}}_\nu\exp(-\mathrm{j}\beta_\nu z)\right)\right] \\ =\mathrm{j}\omega\varepsilon_0\left(\varepsilon_\mathrm{r}+\Delta\varepsilon\right)\sum_\nu A_\nu\vec{\mathcal{E}}_\nu\exp(-\mathrm{j}\beta_\nu z) \tag{C.5}$$

Comparing Eq. (C.4) and (C.5) with Eqs. (B.10) and (B.11), we find

$$\sum_\nu\left[\frac{\partial A_\nu}{\partial z}\vec{e}_z\times\left(\vec{\mathcal{E}}_\nu\exp(-\mathrm{j}\beta_\nu z)\right)\right]=0, \tag{C.6}$$

$$\sum_\nu\left[\frac{\partial A_\nu}{\partial z}\vec{e}_z\times\left(\vec{\mathcal{H}}_\nu\exp(-\mathrm{j}\beta_\nu z)\right)-\mathrm{j}\omega\varepsilon_0\Delta\varepsilon A_\nu\vec{\mathcal{E}}_\nu\exp(-\mathrm{j}\beta_\nu z)\right]=0. \tag{C.7}$$

Multiplying Eq. (C.6) with $\vec{\mathcal{H}}_\mu^*$ and Eq. (C.7) with $\vec{\mathcal{E}}_\mu^*$, and subtracting both equations results in

$$\sum_\nu\frac{\partial A_\nu}{\partial z}\left[\left(\vec{e}_z\times\vec{\mathcal{E}}_\nu\right)\cdot\vec{\mathcal{H}}_\mu^*-\left(\vec{e}_z\times\vec{\mathcal{H}}_\nu\right)\cdot\vec{\mathcal{E}}_\mu^*\right]\exp(-\mathrm{j}\beta_\nu z) \\ =-\mathrm{j}\omega\varepsilon_0\Delta\varepsilon\sum_\nu A_\nu\vec{\mathcal{E}}_\nu\cdot\vec{\mathcal{E}}_\mu^*\exp(-\mathrm{j}\beta_\nu z). \tag{C.8}$$

Making use of the property $\left(\vec{a}\times\vec{b}\right)\cdot\vec{c}=\vec{a}\cdot\left(\vec{b}\times\vec{c}\right)$ and integrating over the cross-section leads to

$$\sum_\nu\iint_{-\infty}^{\infty}\frac{\partial A_\nu}{\partial z}\left[\left(\vec{\mathcal{E}}_\nu\times\vec{\mathcal{H}}_\mu^*\right)+\left(\vec{\mathcal{E}}_\mu^*\times\vec{\mathcal{H}}_\nu\right)\right]\cdot\vec{e}_z\,\mathrm{d}x\mathrm{d}y\cdot\exp(-\mathrm{j}\beta_\mu z) \\ =-\mathrm{j}\omega\varepsilon_0\sum_\nu\iint_{-\infty}^{\infty}\Delta\varepsilon A_\nu\vec{\mathcal{E}}_\nu\cdot\vec{\mathcal{E}}_\mu^*\exp(-\mathrm{j}\beta_\nu z)\mathrm{d}x\mathrm{d}y, \tag{C.9}$$

which can be simplified by making use of the orthogonality relation, Eq. (B.9),

$$\frac{\partial A_\mu}{\partial z}4\mathcal{P}_\mu=-\mathrm{j}\omega\varepsilon_0\sum_\nu\iint_{-\infty}^{\infty}\Delta\varepsilon\vec{\mathcal{E}}_\nu\cdot\vec{\mathcal{E}}_\mu^*\,\mathrm{d}x\mathrm{d}y A_\nu\exp\left[-\mathrm{j}(\beta_\nu-\beta_\mu)z\right]. \tag{C.10}$$

This equation can be rewritten as

$$\frac{\partial A_\mu}{\partial z}=-\mathrm{j}\sum_\nu\kappa_{\mu\nu}A_\nu\exp\left[-\mathrm{j}(\beta_\nu-\beta_\mu)z\right], \tag{C.11}$$

where the so-called coupling coefficient $\kappa_{\mu\nu}$ is given by

$$\kappa_{\mu\nu}=\frac{\omega\varepsilon_0}{4\mathcal{P}_\mu}\iint_{-\infty}^{\infty}\Delta\varepsilon(x,y)\vec{\mathcal{E}}_\nu(x,y)\cdot\vec{\mathcal{E}}_\mu^*(x,y)\mathrm{d}x\mathrm{d}y. \tag{C.12}$$

A non-vanishing coupling coefficient results in an energy transfer between modes. We specialize Eq. (C.11) to the case of self-coupling,

$$\frac{\partial A_\mu(z)}{\partial z} = -\mathrm{j}\kappa_{\mu\mu} A_\mu(z)\,. \tag{C.13}$$

For $\kappa_{\mu\mu} = const$ his differential equation is solved by

$$A_\mu(z) = A\exp(-\mathrm{j}\kappa_{\mu\mu} z)\,. \tag{C.14}$$

Substituting Eq. (C.14) into Eq. (B.7) leads to

$$\vec{E}_\mu(\vec{r},t) = A\vec{\mathcal{E}}_\mu(x,y)\exp\left[\mathrm{j}\left(\omega_\mathrm{c} t - \left(\beta_\mu + \kappa_{\mu\mu}\right)z\right)\right]. \tag{C.15}$$

As expected, a z-independent change in permittivity $\Delta\varepsilon$ leads to a change in the propagation constant $\delta\beta = \kappa_{\mu\mu}$

$$\delta\beta = \kappa_{\mu\mu} = \frac{\omega\varepsilon_0}{4\mathcal{P}_\mu}\iint_{-\infty}^{\infty} \Delta\varepsilon(x,y)\vec{\mathcal{E}}_\mu \cdot \vec{\mathcal{E}}_\mu^* \,\mathrm{d}x\,\mathrm{d}y\,. \tag{C.16}$$

In an SOH modulator this change of permittivity is due to the Pockels effect. Substituting $\Delta\varepsilon$ in Eq. (C.16) by Eq. (2.19) results in an expression for the change of the wave vector due to the Pockels effect, which itself depends on the EO coefficient r_{33} and on the modulation field E_m,

$$\delta\beta = \frac{-\omega_c\varepsilon_0}{4\mathcal{P}_\mu}\iint_{-\infty}^{\infty} n_0^4(x,y) r_{33}(x,y) E_\mathrm{m}(x,y)\,\vec{\mathcal{E}}_\mu(x,y)\cdot\vec{\mathcal{E}}_\mu^*(x,y)\,\mathrm{d}x\,\mathrm{d}y\,. \tag{C.17}$$

Appendix D Electrical Transmission Line

The following section is based on the book "Microwave Engineering" of D. M. Pozar [93]. The reader is referred to Ref. [93] for more information on microwave theory and circuit design. In this section, we will restrict ourselves to the description of a lumped-element equivalent circuit model of a transmission line, and on a basic model for microwave attenuation.

D.1 Lumped-Element Circuit Model

The lumped-element equivalent circuit of a transmission line is depicted in Fig. D.1 and represents a differentially short element of the transmission line of length Δz. The circuit consists of four lumped elements:

R' = Series resistance per length, for both conductors, in [Ω/m],

L' = Series inductance per length, for both conductors, in [H/m],

G' = Shunt conductance per length, in [S/m],

C' = Shunt capacitance per length, in [F/m].

Applying Kirchhoff's voltage and current laws to the equivalent circuit leads to the following expressions [93] for the current $i(z)$ and the voltage $u(z)$ of the transmission line if the source signal is a sinusoidal,

$$\frac{\mathrm{d}u(z)}{\mathrm{d}z} = -(R' + \mathrm{j}\omega_{\mathrm{m}} L')i(z), \tag{D.1}$$

$$\frac{\mathrm{d}i(z)}{\mathrm{d}z} = -(G' + \mathrm{j}\omega_{\mathrm{m}} LC')u(z). \tag{D.2}$$

Cross-substituting in Eq. (D.1) and Eq. (D.2) results in the wave equations for voltage and current,

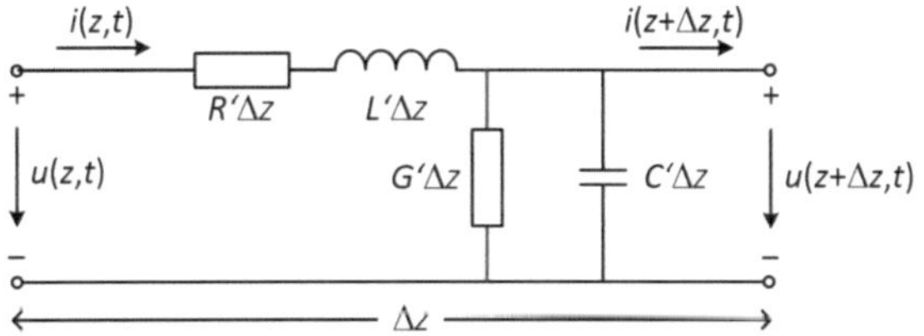

Fig. D.1 Lumped-element equivalent circuit of a transmission line. Redrawn after [93]

$$\frac{\mathrm{d}^2 u(z)}{\mathrm{d}z^2} - \gamma^2 u(z) = 0, \tag{D.3}$$

$$\frac{\mathrm{d}^2 i(z)}{\mathrm{d}z^2} - \gamma^2 i(z) = 0, \tag{D.4}$$

where the complex propagation constant γ is given by

$$\gamma = \frac{\alpha_\mathrm{m}}{2} + \mathrm{j}\beta = \sqrt{(R' + \mathrm{j}\omega_\mathrm{m} L')(G' + \mathrm{j}\omega_\mathrm{m} C')}\,. \tag{D.5}$$

As before, the quantity α_m denotes the power attenuation coefficient, and β is the propagation constant.

The wave equations (D.3) and (D.4) can be solved by the ansatz

$$u(z) = u_0^+ e^{-\gamma z} + u_0^- e^{\gamma z}, \tag{D.6}$$

$$i(z) = i_0^+ e^{-\gamma z} + i_0^- e^{\gamma z}. \tag{D.7}$$

Substituting Eq. (D.6) into Eq. (D.1) leads to

$$i(z) = \frac{\gamma}{R' + \mathrm{j}\omega_\mathrm{m} L'}\left[u_0^+ e^{-\gamma z} - u_0^- e^{\gamma z}\right]. \tag{D.8}$$

Comparing Eq. (D.8) and (D.7) leads to the definition of the line impedance Z_L,

$$Z_\mathrm{L} = \frac{u_0^+}{i_0^+} = -\frac{u_0^-}{i_0^-} = \frac{R' + \mathrm{j}\omega_\mathrm{m} L'}{\gamma} = \frac{\sqrt{R' + \mathrm{j}\omega_\mathrm{m} L'}}{\sqrt{G' + \mathrm{j}\omega_\mathrm{m} C'}}\,. \tag{D.9}$$

D.2 Lossy Transmission Line

From the complex propagation constant, Eq. (D.5), we see that the transmission line parameters R' and G' are responsible for attenuation. The series resistance R' represents conductor loss, whereas the shunt conductance G' stands for dielectric loss. For the case of a low-loss transmission line we may simplify the expression for the complex propagation constant γ by assuming $R' \ll \omega_\mathrm{m} L'$ and $G' \ll \omega_\mathrm{m} C'$,

$$\gamma = \sqrt{(\mathrm{j}\omega_\mathrm{m}L')(\mathrm{j}\omega_\mathrm{m}C')\left(1+\frac{R'}{\mathrm{j}\omega_\mathrm{m}L'}\right)\left(1+\frac{G'}{\mathrm{j}\omega_\mathrm{m}C'}\right)}$$
$$\approx \mathrm{j}\omega_\mathrm{m}\sqrt{L'C'}\sqrt{1-\mathrm{j}\left(\frac{R'}{\omega_\mathrm{m}L'}+\frac{G'}{\omega_\mathrm{m}C'}\right)}. \tag{D.10}$$

Performing a first-order Taylor series expansion $\sqrt{1+x} \approx 1 + x/2$ we find

$$\gamma \approx \mathrm{j}\omega_\mathrm{m}\sqrt{L'C'}\left[1-\frac{\mathrm{j}}{2}\left(\underbrace{\frac{R'}{\omega_\mathrm{m}L'}}_{\tan\delta_\mathrm{L}}+\underbrace{\frac{G'}{\omega_\mathrm{m}C'}}_{\tan\delta_\mathrm{C}}\right)\right], \tag{D.11}$$

and identify propagation constant β, phase velocity v, and power attenuation constant α,

$$\beta \approx \omega_\mathrm{m}\sqrt{L'C'}, \qquad v = \frac{1}{\sqrt{L'C'}}, \tag{D.12}$$

$$\alpha_\mathrm{m} \approx \underbrace{R'\sqrt{\frac{C'}{L'}}}_{\text{conductor loss}} + \underbrace{G'\sqrt{\frac{L'}{C'}}}_{\text{dielectric loss}} \approx \frac{R'}{Z_\mathrm{L}} + G'Z_\mathrm{L}. \tag{D.13}$$

The quantities $\tan\delta_\mathrm{L}$ and $\tan\delta_\mathrm{C}$ in Eq. (D.11) are called loss tangents. In Eq. (D.13) the line impedance was approximated by $Z_\mathrm{L} \approx \sqrt{L'/C'}$.

We will now investigate the frequency-dependency of conductor loss and dielectric loss separately. The frequency dependency of the conductor loss has its origin in the skin effect: In a conductor with conductivity σ the complex electric permittivity ε_r-jε_i and the complex propagation constant γ follow from Maxwell's equations ([93], Eq. (1.50a,b))

$$\varepsilon_\mathrm{r} - \mathrm{j}\varepsilon_\mathrm{i} = \varepsilon_\mathrm{r}\left(1-\mathrm{j}\frac{\sigma}{\varepsilon_0\varepsilon_\mathrm{r}\omega_\mathrm{m}}\right)$$
$$\gamma = \frac{\alpha_\mathrm{m}}{2} + \mathrm{j}\beta = \mathrm{j}\omega_\mathrm{m}\sqrt{\varepsilon_0\mu_0}\sqrt{\mu_\mathrm{r}}\sqrt{\varepsilon_\mathrm{r}-\mathrm{j}\varepsilon_\mathrm{i}} = \mathrm{j}\omega_\mathrm{m}\sqrt{\varepsilon_0\varepsilon_\mathrm{r}\mu_0\mu_\mathrm{r}}\sqrt{1-\mathrm{j}\frac{\sigma}{\varepsilon_0\varepsilon_\mathrm{r}\omega_\mathrm{m}}}. \tag{D.14}$$

Here, μ_r and ε_r are the relative permeability and the relative permittivity of the conductor, respectively. For a good conductor $\sigma \gg \varepsilon_0\varepsilon_\mathrm{r}\omega_\mathrm{m}$ is fulfilled, and we can approximate the propagation constant by

$$\gamma \approx \mathrm{j}\omega_\mathrm{m}\sqrt{\varepsilon_0\varepsilon_\mathrm{r}\mu_0\mu_\mathrm{r}}\sqrt{-\mathrm{j}\frac{\sigma}{\varepsilon_0\varepsilon_\mathrm{r}\omega_\mathrm{m}}} = \frac{1+\mathrm{j}}{\sqrt{2}}\sqrt{\omega_\mathrm{m}\mu_0\mu_\mathrm{r}\sigma}. \tag{D.15}$$

The imaginary part of the propagation constant results in an attenuation of the electric field. After a penetration depth $\delta_s = 2/\alpha_m = \sqrt{2/(\omega_m \mu_0 \mu_r \sigma)}$, also known as skin depth, the field amplitude assumes the fraction $1/e$ from its maximum. This means that the corresponding alternating current only flows in a thin conducting channel of thickness $\delta_s \propto 1/\sqrt{\omega_m}$ at the surface of the conductor. The frequency-dependency of the conductor power loss constant is

$$\alpha_{R'} = \frac{2}{\delta_s} = \sqrt{2\mu_0\mu_r\sigma}\sqrt{\omega_m} \approx \frac{R'}{Z_L}. \tag{D.16}$$

Conductor loss thus scales with the square root of the angular frequency ω_m. The expression for the dielectric power loss constant in Eq. (D.13) can be written as

$$\alpha_{G'} \approx G'Z_L = \omega_m C' \tan\delta_C \sqrt{\frac{L'}{C'}} = \beta \tan\delta_C = \frac{\omega_m}{v}\tan\delta_C. \tag{D.17}$$

In the case of a good insulator and for frequencies far away from any absorption resonance of the dielectric, the loss angle is small $\tan\delta_C = \varepsilon_i / \varepsilon_r \ll 1$ and can be considered to be frequency independent, $\varepsilon_i(\omega_m)/\varepsilon_r(\omega_m) \approx \text{const}$. In this case we find that the dielectric loss increases linearly with the angular frequency $\alpha_{G'} \propto \omega_m$. An intuitive explanation for this behavior is that dielectric loss is proportional to the occurrence of polarization reversals (and thus to the frequency).

Appendix E Mach-Zehnder Modulator

In the following, we derive the operation principle of an MZM that relies on phase shifting elements. An MZM consists of a Mach-Zehnder interferometer (MZI) that comprises two couplers with an amplitude coupling ratio of $1/\sqrt{2}$, and one or more phase-shifters in the arms of the MZI. The fields at the two output ports of the couplers have a phase difference of $\pi/2$. In general, the MZM can be represented by two dynamic phase shifter sections φ_1, φ_2 and a static bias phase shifter section φ_{bias}, as depicted in Fig. E.1(a). In the case of Pockels effect phase shifters the phase shift is proportional to an applied voltage $\delta\varphi_{1/2} \propto U_{\text{ext}}$, see Eq. (2.36). The MZM usually has two input and two output ports. The field amplitude transfer functions of a cross transition ($E_{\text{in}} \rightarrow E_{\text{out,1}}$) and of a bar transition ($E_{\text{in}} \rightarrow E_{\text{out,2}}$) are given by the following matrix products

$$E_{\text{out,1}} = \begin{pmatrix} 1/\sqrt{2} & e^{j\pi/2}/\sqrt{2} \end{pmatrix} \begin{pmatrix} e^{j(\varphi_1+\varphi_{\text{bias}})} & 0 \\ 0 & e^{j\varphi_2} \end{pmatrix} \begin{pmatrix} e^{j\pi/2}/\sqrt{2} \\ 1/\sqrt{2} \end{pmatrix} E_{\text{in}}, \tag{E.1}$$

$$E_{\text{out,2}} = \begin{pmatrix} e^{j\pi/2}/\sqrt{2} & 1/\sqrt{2} \end{pmatrix} \begin{pmatrix} e^{j(\varphi_1+\varphi_{\text{bias}})} & 0 \\ 0 & e^{j\varphi_2} \end{pmatrix} \begin{pmatrix} e^{j\pi/2}/\sqrt{2} \\ 1/\sqrt{2} \end{pmatrix} E_{\text{in}}, \tag{E.2}$$

that can be rewritten as

$$E_{\text{out,1}} = \frac{1}{2} E_{\text{in}} e^{j(\varphi_1+\varphi_2+\varphi_{\text{bias}}+\pi)/2)} \left(e^{j(\varphi_1-\varphi_2+\varphi_{\text{bias}})/2} + e^{-j(\varphi_1-\varphi_2+\varphi_{\text{bias}})/2} \right), \tag{E.3}$$

$$E_{\text{out,2}} = \frac{1}{2j} E_{\text{in}} e^{j(\varphi_1+\varphi_2+\varphi_{\text{bias}}+\pi)/2)} \left(-e^{j(\varphi_1-\varphi_2+\varphi_{\text{bias}})/2} + e^{-j(\varphi_1-\varphi_2+\varphi_{\text{bias}})/2} \right). \tag{E.4}$$

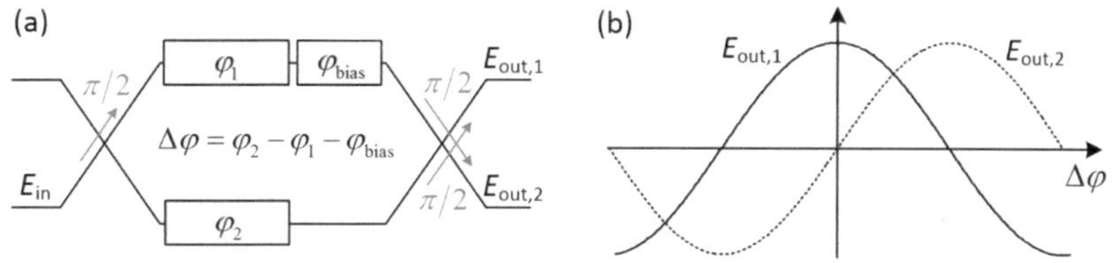

Fig. E.1 Operation principle of a Mach-Zehnder modulator. (a) Schematic of an MZM. Two directional couplers are used as power dividers/combiners. Up to three phase shifters are used to adjust the relative phase difference φ of the two arms of the MZI. (b) Transfer functions of the MZM.

With the phase difference $\Delta\varphi = \varphi_2 - \varphi_1 - \varphi_{\text{bias}}$ we rewrite Eqs. (E.3) and (E.4)

$$E_{\text{out},1} = E_{\text{in}} e^{\mathrm{j}(\varphi_1+\varphi_2+\varphi_{\text{bias}}+\pi)/2)} \quad \cos\left(\frac{\Delta\varphi}{2}\right), \tag{E.5}$$

$$E_{\text{out},2} = E_{\text{in}} \underbrace{e^{\mathrm{j}(\varphi_1+\varphi_2+\varphi_{\text{bias}}+\pi)/2)}}_{\text{Phase modulation}} \quad \underbrace{\sin\left(\frac{\Delta\varphi}{2}\right)}_{\text{Amplitude modulation}}. \tag{E.6}$$

The transfer functions thus consist of a phase modulation term and of an amplitude modulation term.

E.1 Push-Pull Operation

The existence of simultaneous amplitude and phase modulation is usually disadvantageous, since phase modulation results in a chirp that can enhance signal distortion during transmission through a dispersive medium. Detrimental phase modulation of an MZM can be eliminated by opposite phase shifts $\varphi_1 = -\varphi_2$ in the two arms of the MZM, such that $\Delta\varphi = 2\varphi_2 - \varphi_{\text{bias}}$

$$E_{\text{out},1} = E_{\text{in}} e^{\mathrm{j}(\varphi_{\text{bias}}+\pi)/2} \cdot \cos\left(\varphi_2 - \frac{\varphi_{\text{bias}}}{2}\right), \tag{E.7}$$

$$E_{\text{out},2} = E_{\text{in}} \underbrace{e^{\mathrm{j}(\varphi_{\text{bias}}+\pi)/2}}_{\text{Static}} \cdot \underbrace{\sin\left(\varphi_2 - \frac{\varphi_{\text{bias}}}{2}\right)}_{\text{Amplitude modulation}}. \tag{E.8}$$

This mode of driving the modulator is called push-pull operation.

E.2 On-Off Keying (OOK)

For OOK the modulator switches between the state of maximum transmission ("on") and the state of minimum transmission ("off"). Assuming push-pull operation the intensity transfer function is given by

$$I_{\text{out},1} = I_{\text{in}} \cos^2\left(\varphi_2 - \frac{\varphi_{\text{bias}}}{2}\right) = \frac{I_{\text{in}}}{2}\left[1 + \cos\left(\underbrace{2\varphi_2}_{\varphi_2-\varphi_1} - \varphi_{\text{bias}}\right)\right]. \tag{E.9}$$

Thus a phase difference of $\varphi_2 - \varphi_1 = 2\varphi_2 = \pi$ is required to switch between the state of maximum transmission to the state of minimum transmission. The corresponding drive voltage is called π-voltage U_π. At the so-called quadrature point the bias is adjusted such that the output intensity is half of its maximum value.

E.3 Bipolar Amplitude-Shift Keying (ASK)

The data rate of an MZM can be increased while maintaining the symbol rate by encoding more than one bit per symbol. This can be done by bipolar *M*-ASK, where *M* symbol states are generated using *M*/2 amplitude levels and 2 phase states (0°, 180°). The case of bipolar 4ASK is depicted in Fig. E.2. Here, 4 equidistant amplitude levels (4 symbol states, 2 bits) are used to drive the MZM. The operation point of the MZM is adjusted to zero transmission. The drive amplitudes are chosen such that the MZM is operated in the linear regime of its field transfer function. The corresponding constellation diagram of the bipolar 4ASK signal thus comprises 4 equidistant symbols. The use of bipolar 4ASK allows to double the data rate at a given symbol rate as compared to OOK or bipolar 2ASK (also called "binary phase shift keying" - BPSK) modulation. However, a coherent receiver is required to measure both amplitude and phase of the optical signal, see Appendix F.2.

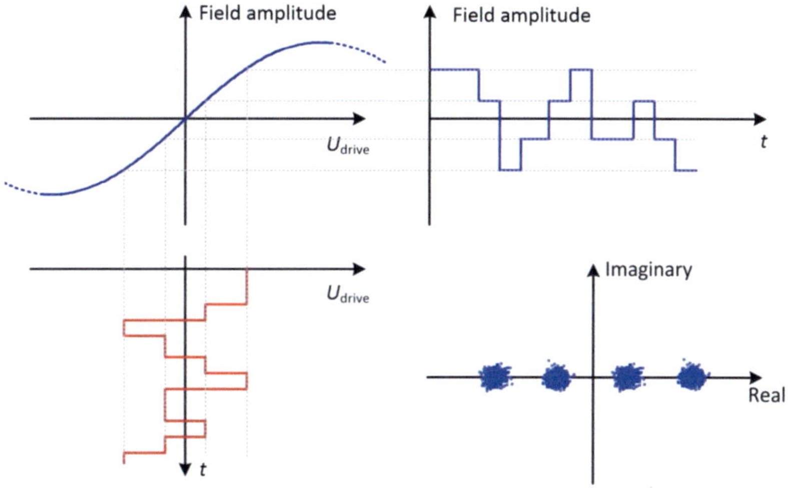

Fig. E.2 Bipolar amplitude shift keying using a push-pull MZM. Equidistant drive voltage amplitudes (red) are chosen such that the modulator is driven in the linear regime of its transfer function. The operation point is adjusted to zero transmission by applying a bias voltage. This way a 4-level electrical signal results in an optical 4ASK signal.

Appendix F Signal Reception and Analysis

Depending on the modulation format different signal reception schemes need to be utilized.

F.1 Direct Detection

In the case of intensity modulation, such as OOK, the signal can be directly detected by a single photodiode (PD). The signal intensity $I(t)$ can be split into a time-dependent part $\delta I(t)$ and into a constant part I_c

$$I(t) = I_c + \delta I(t). \tag{F.1}$$

The photocurrent i_{PD} of the PD is directly proportional to the intensity of the incident light

$$i_{PD} \propto I_c + \delta I(t). \tag{F.2}$$

The PD is connected to the 50Ω input impedance of a high-speed oscilloscope – usually a versatile digital communications analyzer (DCA). The intensity modulated optical signal is then measured as PD voltage at the DCA.

F.2 Coherent Reception

When using complex modulation formats not only the amplitude of each symbol but also the corresponding phase needs to be resolved, which cannot be achieved by direct detection. Therefore a so-called coherent receiver is required. For coherent reception the signal is mixed with a second, non-modulated optical carrier, the local oscillator (LO). The reception is called homodyne if the carrier frequencies of the LO and signal are identical, and it is called intradyne if the LO frequency differs from the signal frequency, but still lies within the signal bandwidth. A schematic of a coherent receiver is depicted in Fig. F.1. Using four 2×2 MMI couplers (output fields feature 90° phase difference) the electric fields E_s and E_{LO} of signal and local oscillator are mixed, such that the electric fields at the inputs of the balanced PDs are given as

$$\begin{aligned}
E_1(t) &= [E_s(t) - E_{LO}(t)]/2, \\
E_2(t) &= \mathrm{j}[E_s(t) + E_{LO}(t)]/2, \\
E_3(t) &= [\mathrm{j}E_s(t) - E_{LO}(t)]/2, \\
E_4(t) &= [E_{LO}(t) - \mathrm{j}E_s(t)]/2.
\end{aligned} \tag{F.3}$$

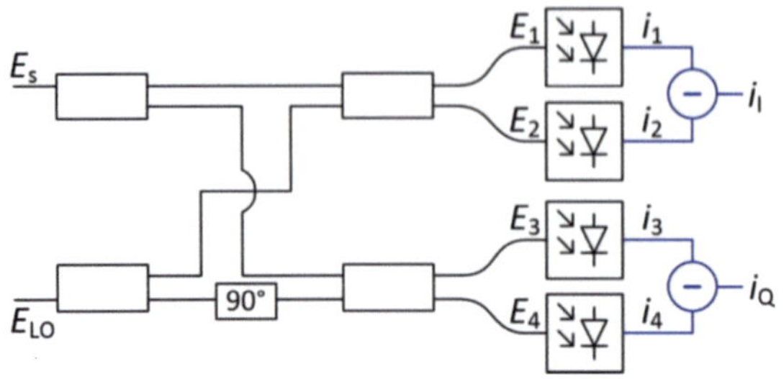

Fig. F.1 Schematic of a coherent receiver that uses four 2×2 MMI couplers (rectangles) and two pairs of balanced PDs. Black lines depict optical WGs, blue lines depict electrical connections. All optical paths from E_s, E_{LO} to the PDs are matched in length.

At the four PDs the currents i_1, i_2, i_3, i_4 are generated. Subtracting these currents pairwise results in two currents i_I and i_Q that are proportional to the in-phase and quadrature-phase component of the signal amplitude

$$\begin{aligned} i_I(t) &= i_2(t) - i_1(t) \propto \mathrm{Re}\left\{E_s(t)E_{LO}^*(t)\right\}, \\ i_Q(t) &= i_4(t) - i_3(t) \propto \mathrm{Im}\left\{E_s(t)E_{LO}^*(t)\right\}. \end{aligned} \tag{F.4}$$

F.3 Quality Metrics for Optical Signals

Optical signal quality needs to be sufficiently high to enable error-free reception at the end of a transmission link. In practice, a small number of errors can be accepted by applying forward error-correction (FEC) algorithms, e.g., a bit error ratio (BER) below 4.5×10^{-3} can be tolerated when using the latest hard-decision FEC algorithms [144].

For very high signal qualities counting of bit errors becomes impractical due to the required measurement time. In that case other quality metrics can be used to predict the associated BER. In the case of intensity modulated OOK signals a commonly used metric is the so-called quality factor, or Q-factor. This factor can directly be determined when using a digital communications analyzer. The Q-factor can be used to calculate the BER of a signal if the following three assumptions are fulfilled [157]: Equal probability of ones and zeros, optimum decision threshold, Gaussian probability distribution for ones and zeros (i. e., white Gaussian noise). The probability $P_x(u)$ for measuring a voltage u at a certain decision instant is thus given by

$$P_x(u) = \frac{1}{\sqrt{2\pi\sigma_x^2}} \exp\left[\frac{(u - u_x)^2}{2\sigma_x^2}\right], \tag{F.5}$$

where u_x and σ_x are the mean value and the standard deviation of a symbol state x. The mean voltage u_0 refers to the ensemble of detected zeros that has a value of 0 in an ideal case, u_1 refers to the ensemble of received ones. Under these assumptions the linear quality factor Q is defined as

$$Q = \frac{u_1 - u_0}{\sigma_1 + \sigma_0}. \tag{F.6}$$

The BER is then linked to the quality factor by [158]

$$\text{BER} = \frac{1}{2}\text{erfc}\left(\frac{Q}{\sqrt{2}}\right), \tag{F.7}$$

where erfc(z) is the complementary error-function

$$\text{erfc}(z) = \frac{2}{\sqrt{\pi}}\int_z^\infty \exp\left(-(z')^2\right)\mathrm{d}z'. \tag{F.8}$$

Usually the Q-factor is either denoted in linear units, or as Q^2-factor in logarithmic units

$$Q^2\Big|_{\text{dB}} = 20\log_{10}(Q). \tag{F.9}$$

For complex modulation formats like QAM the so-called error vector magnitude (EVM) [111] is used as a quality metric. Here, the received complex electric field $\vec{E}_{\text{r},i}$ of a symbol i is compared to its targeted electric field $\vec{E}_{\text{t},i}$. The error vector $\vec{E}_{\text{err},i}$ is used to describe the deviation between received and targeted field

$$\vec{E}_{\text{err},i} = \vec{E}_{\text{r},i} - \vec{E}_{\text{t},i}. \tag{F.10}$$

Given a certain sample size N, the variance σ_{err}^2 of the measurement can be obtained

$$\sigma_{\text{err}}^2 = \frac{1}{N}\sum_{i=1}^{N}\left|\vec{E}_{\text{err},i}\right|^2. \tag{F.11}$$

By normalizing the quantity σ_{err} by the maximum targeted field amplitude $\left|\vec{E}_{\mathrm{t,m}}\right|$ the so-called error vector magnitude with maximum normalization $\mathrm{EVM_m}$ is obtained

$$\mathrm{EVM_m} = \frac{\sigma_{\mathrm{err}}}{\left|\vec{E}_{\mathrm{t,m}}\right|}. \tag{F.12}$$

According to [111] the EVM of an M-QAM signal is linked to its corresponding BER, if additive white Gaussian noise, and non-data-aided reception can be assumed:

$$\mathrm{BER} \approx \frac{\left(1-m^{-1}\right)}{\log_2 m}\mathrm{erfc}\left(\sqrt{\frac{3\log_2 m}{(m^2-1)(k\,\mathrm{EVM_m})^2\log_2 M}}\right). \tag{F.13}$$

Here, M denotes the number of symbol states, m is the number of signal levels per dimension of the constellation and k is defined as ratio between targeted maximum field amplitude $\left|\vec{E}_{\mathrm{t,m}}\right|$ and targeted average field amplitude $\left|\vec{E}_{\mathrm{t,a}}\right|$

$$k = \frac{\left|\vec{E}_{\mathrm{t,m}}\right|}{\left|\vec{E}_{\mathrm{t,a}}\right|}. \tag{F.14}$$

For exemplary modulation formats k^2 takes the form:

	QPSK	16QAM	32QAM	64QAM
k^2	1	9/5	17/10	7/3

Appendix G Dispersion and Absorption of EO Organic Materials

In the following the measured complex refractive index of two investigated EO materials as a function of wavelength is given. Measurements have been carried out by Delwin Elder, University of Washington, Seattle, using ellipsometry.

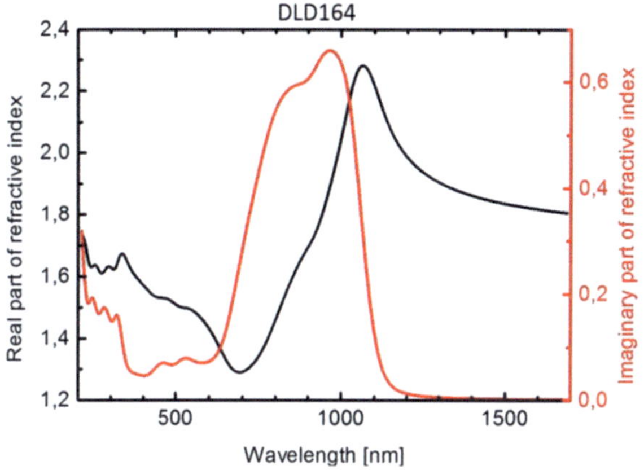

Fig. G.1 Complex refractive index of the structurally engineered material DLD164. At a wavelength of 1550 nm the real part is 1.83 and the imaginary part is 0.002.

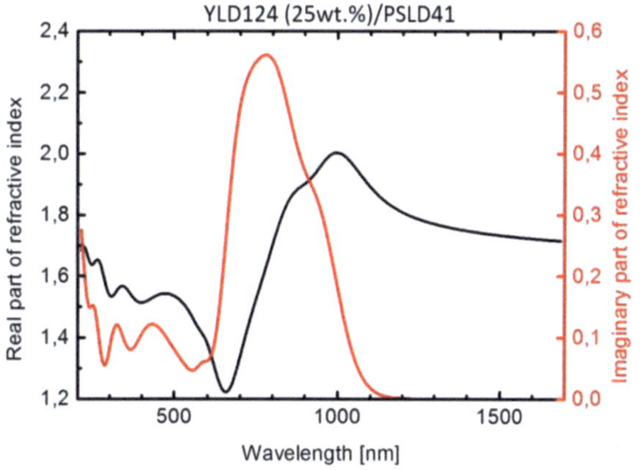

Fig. G.2 Complex refractive index of the binary chromophores organic glass composed of a mixture of 25wt.% YLD124 and PSLD41. At a wavelength of 1550 nm the real part is 1.73 and the imaginary part is 0.000001.

Appendix H Bibliography

[1] "Datacenterknowledge.com" [Online] Available: http://www.datacenterknowledge.com/inside-facebooks-lulea-data-center/

[2] "Cisco Global Cloud Index: Forecast and Methodology, 2012-2017." [Online]. Available: http://www.cisco.com/c/en/solutions/collateral/service-provider/global-cloud-index-gci/Cloud_Index_White_Paper.pdf.

[3] S. K. Selvaraja, W. Bogaerts, P. Absil, D. Van Thourhout, and R. Baets, "Record Low-Loss Hybrid Rib/Wire Waveguides for Silicon Photonic Circuits," in *Group IV Photonics, 7th International conference*, 2010.

[4] S. Selvaraja, M. Gayle, M. Alexey, D. Christie, O. Patrick, S. Pathak, D. Vermeulen, G. Sterckx, G. Winroth, P. Verheyen, G. Lepage, R. Baets, W. Bogaerts, J. Van Campenhout, and P. Absil, "Advanced 300-mm Waferscale Patterning for Silicon Photonics Devices with Record Low Loss and Phase Errors," in *17th OptoElectronics and Communications Conference (OECC 2012)*, 2012, pp. PDP2–2.

[5] S.-C. Hung, E.-Z. Liang, and C.-F. Lin, "Silicon Waveguide Sidewall Smoothing by KrF Excimer Laser Reformation," *J. Light. Technol.*, vol. 27, no. 7, pp. 887–892, Apr. 2009.

[6] C. G. Poulton, C. Koos, M. Fujii, A. Pfrang, T. Schimmel, J. Leuthold, and W. Freude, "Radiation Modes and Roughness Loss in High Index-Contrast Waveguides," *Sel. Top. Quantum Electron. IEEE J.*, vol. 12, no. 6, pp. 1306–1321, 2006.

[7] U. Fischer, T. Zinke, J. Kropp, F. Arndt, and K. Petermann, "0.1 dB/cm waveguide losses in single-mode SOI rib waveguides," *Photonics Technol. Lett. IEEE*, vol. 8, no. 5, pp. 647–648, 1996.

[8] J. Cardenas, C. B. Poitras, J. T. Robinson, K. Preston, L. Chen, and M. Lipson, "Low loss etchless silicon photonic waveguides," *Opt. Express*, vol. 17, no. 6, pp. 4752–4757, 2009.

[9] X. Jin-Song, Y. Jin-Zhong, F. Zhong-Chao, W. Zhang-Tao, and C. Shao-Wu, "Multimode Interference 3-dB Coupler in Silicon-on-Insulator Based on Silicon Rib Waveguides with Trapezoidal Cross Section," *Chinese Phys. Lett.*, vol. 21, no. 1, pp. 104–106, Jan. 2004.

[10] J. V Galán, P. Sanchis, G. Sánchez, and J. Mart'i, "Polarization insensitive low-loss coupling technique between SOI waveguides and high mode field diameter single-mode fibers," *Opt. Express*, vol. 15, no. 11, pp. 7058–7065, 2007.

[11] Q. Fang, J. Song, X. Luo, M. Yu, G. Lo, and Y. Liu, "Mode-size converter with high coupling efficiency and broad bandwidth," *Opt. Express*, vol. 19, no. 22, pp. 21588–21594, 2011.

[12] N. Lindenmann, G. Balthasar, D. Hillerkuss, R. Schmogrow, M. Jordan, J. Leuthold, W. Freude, and C. Koos, "Photonic wire bonding: a novel concept for chip-scale interconnects," *Opt. Express*, vol. 20, no. 16, pp. 17667–17677, Jul. 2012.

[13] W. S. Zaoui, A. Kunze, W. Vogel, M. Berroth, J. Butschke, and F. Letzkus, "CMOS-compatible nonuniform grating coupler with 86% coupling efficiency," in *39th European Conference and Exhibition on Optical Communication (ECOC 2013),* 2013.

[14] B. Snyder and P. O'Brien, "Planar Fiber Packaging Method for Silicon Photonic Integrated Circuits," in *Optical Fiber Communication Conference*, 2012, p. OM2E.5.

[15] R. S. Jacobsen, K. N. Andersen, P. I. Borel, J. Fage-Pedersen, L. H. Frandsen, O. Hansen, M. Kristensen, A. V Lavrinenko, G. Moulin, H. Ou, C. Peucheret, B. Zsigri, and A. Bjarklev, "Strained silicon as a new electro-optic material," *Nature*, vol. 441, pp. 199–202, 2006.

[16] B. Chmielak, M. Waldow, C. Matheisen, C. Ripperda, J. Bolten, T. Wahlbrink, M. Nagel, F. Merget, and H. Kurz, "Pockels effect based fully integrated, strained silicon electro-optic modulator," *Opt. Express*, vol. 19(18), pp. 17212–17219, 2011.

[17] G. T. Reed, G. Mashanovich, F. Y. Gardes, and D. J. Thomson, "Silicon optical modulators," *Nat. Phot.*, vol. 4, no. 8, pp. 518–526, Jul. 2010.

[18] W. M. Green, M. J. Rooks, L. Sekaric, and Y. A. Vlasov, "Ultra-compact, low RF power, 10 Gb/s siliconMach-Zehnder modulator," *Opt. Express*, vol. 15, no. 25, pp. 17106–17113, 2007.

[19] T. Baba, S. Akiyama, M. Imai, T. Akagawa, M. Takahashi, N. Hirayama, H. Takahashi, Y. Noguchi, H. Okayama, T. Horikawa, and T. Usuki, "25-Gbps operation of silicon p-i-n Mach-Zehnder optical modulator with 100-µm-long phase shifter," *Conf. Lasers Electro-Optics (CLEO 2012), San Jose*, 2012.

[20] D. J. Thomson, F. Y. Gardes, J. Fedeli, S. Zlatanovic, Y. Hu, B. P. P. Kuo, E. Myslivets, N. Alic, S. Radic, G. Z. Mashanovich, and G. T. Reed, "50-Gb/s Silicon Optical Modulator," *IEEE Photonics Technol. Lett.*, vol. 24, no. 4, pp. 234–236, Feb. 2012.

[21] P. Dong, C. Xie, L. Chen, L. L. Buhl, and Y. Chen, "112-Gb/s Monolithic PDM-QPSK Modulator in Silicon," in *European Conference on Optical Communication (ECOC 2012),* PDP Paper Th.3.B.1, 2012.

[22] P. Dong, X. Liu, C. Sethumadhavan, L. L. Buhl, R. Aroca, Y. Baeyens, and Y.-K. Chen, "224-Gb/s PDM-16-QAM Modulator and Receiver based on Silicon Photonic Integrated Circuits," in *Optical Fiber Communication Conference/National Fiber Optic Engineers Conference (OFC 2013)*, PDP Paper PDP5C.6, 2013.

[23] T. Baehr-Jones, R. Ding, Y. Liu, A. Ayazi, T. Pinguet, N. C. Harris, M. Streshinsky, P. Lee, Y. Zhang, A. E.-J. Lim, T.-Y. Liow, S. H.-G. Teo, G.-Q. Lo, and M. Hochberg, "Ultralow drive voltage silicon traveling-wave modulator," *Opt. Express*, vol. 20, no. 11, p. 12014, May 2012.

[24] M. R. Watts, W. a Zortman, D. C. Trotter, R. W. Young, and A. L. Lentine, "Vertical junction silicon microdisk modulators and switches," *Opt. Express*, vol. 19, no. 22, pp. 21989–2003, Oct. 2011.

[25] H.-W. Chen, J. D. Peters, and J. E. Bowers, "Forty Gb/s hybrid silicon Mach-Zehnder modulator with low chirp," *Opt. Express*, vol. 19, no. 2, pp. 1455–1460, 2011.

[26] J. Leuthold, W. Freude, J.-M. Brosi, R. Baets, P. Dumon, I. Biaggio, M. L. Scimeca, F. Diederich, B. Frank, and C. Koos, "Silicon Organic Hybrid Technology-A Platform for Practical Nonlinear Optics," *Proc. IEEE*, vol. 97, no. 7, pp. 1304–1316, Jul. 2009.

[27] V. R. Almeida, Q. F. Xu, C. A. Barrios, and M. Lipson, "Guiding and confining light in void nanostructure," *Opt. Lett.*, vol. 29, no. 11, pp. 1209–1211, Jun. 2004.

[28] L. R. Dalton, D. Lao, B. C. Olbricht, S. Benight, D. H. Bale, J. A. Davies, T. Ewy, S. R. Hammond, and P. a. Sullivan, "Theory-inspired development of new nonlinear optical materials and their integration into silicon photonic circuits and devices," *Opt. Mater. (Amst).*, vol. 32, no. 6, pp. 658–668, Apr. 2010.

[29] J.-M. Brosi, C. Koos, L. C. Andreani, M. Waldow, J. Leuthold, and W. Freude, "High-speed low-voltage electro-optic modulator with a polymer-infiltrated silicon photonic crystal waveguide," *Opt. Express*, vol. 16, no. 6, pp. 4177–4191, 2008.

[30] J. H. Wuelbern, S. Prorok, J. Hampe, A. Petrov, M. Eich, J. Luo, A. K.-Y. Jen, M. Jenett, and A. Jacob, "40 GHz electro-optic modulation in hybrid silicon-organic slotted photonic crystal waveguides," *Opt. Lett.*, vol. 35, no. 16, pp. 2753–2755, Aug. 2010.

[31] L. Alloatti, D. Korn, R. Palmer, D. Hillerkuss, J. Li, A. Barklund, R. Dinu, J. Wieland, M. Fournier, J. Fedeli, H. Yu, W. Bogaerts, P. Dumon, R. Baets, C. Koos, W. Freude, and J. Leuthold, "42.7 Gbit/s electro-optic modulator in silicon technology," *Opt. Express*, vol. 19, no. 12, pp. 11841–11851, Jun. 2011.

[32] R. Ding, T. Baehr-Jones, Y. Liu, R. Bojko, J. Witzens, S. Huang, J. Luo, S. Benight, P. Sullivan, J.-M. Fedeli, M. Fournier, L. Dalton, a Jen, and M. Hochberg, "Demonstration of a low V pi L modulator with GHz bandwidth based on electro-optic polymer-clad silicon slot waveguides," *Opt. Express*, vol. 18, no. 15, pp. 15618–15623, 2010.

[33] R. Palmer, L. Alloatti, D. Korn, P. C. Schindler, M. Baier, J. Bolten, T. Wahlbrink, M. Waldow, R. Dinu, W. Freude, C. Koos, and J. Leuthold, "Low Power Mach-Zehnder Modulator in Silicon-Organic Hybrid Technology," *IEEE Photonics Technol. Lett.*, vol. 25, no. 13, pp. 1226–1229, 2013.

[34] R. Palmer, L. Alloatti, D. Korn, P. C. Schindler, R. Schmogrow, W. Heni, S. Koenig, J. Bolten, T. Wahlbrink, M. Waldow, H. Yu, W. Bogaerts, P. Verheyen, G. Lepage, M. Pantouvaki, J. Van Campenhout, P. Absil, R. Dinu, W. Freude, C. Koos, and J. Leuthold, "Silicon-Organic Hybrid MZI Modulator Generating OOK, BPSK and 8-ASK Signals for Up to 84 Gbit/s," *IEEE Photonics J.*, vol. 5, no. 2, pp. 6600907–6600907, Apr. 2013.

[35] D. Korn, R. Palmer, H. Yu, P. C. Schindler, L. Alloatti, M. Baier, R. Schmogrow, W. Bogaerts, S. K. Selvaraja, G. Lepage, M. Pantouvaki, J. M. D. Wouters, P. Verheyen, J. Van Campenhout, B. Chen, R. Baets, P. Absil, R. Dinu, C. Koos, W. Freude, and J. Leuthold, "Silicon-organic hybrid (SOH) IQ modulator using the linear electro-optic effect for transmitting 16QAM at 112 Gbit/s," *Opt. Express*, vol. 21, no. 11, pp. 13219–13227, Jun. 2013.

[36] R. Ding, T. Baehr-Jones, W.-J. Kim, A. Spott, M. Fournier, J.-M. Fedeli, S. Huang, J. Luo, A. K.-Y. Jen, L. Dalton, and M. Hochberg, "Sub-Volt Silicon-Organic Electro-optic Modulator With 500 MHz Bandwidth," *J. Light. Technol.*, vol. 29, no. 8, pp. 1112–1117, Apr. 2011.

[37] R. W. Boyd, *Nonlinear Optics*, 3rd ed. Academic Press, 2008.

[38] J. J. Wolff and R. Wortmann, "Organic Materials for Second-Order Non-Linear Optics," *Adv. Phys. Org. Chem.*, vol. 32, pp. 121–217, 1999.

[39] Jia-ming Liu, *Photonic Devices*. Cambridge University Press, 2005.

[40] J.-M. Brosi, "Slow-Light Photonic Crystal Devices for High-Speed Optical Signal Processing," Doctoral thesis - University of Karlsruhe, 2008.

[41] S. Koeber, "Characterization of photorefractive polymer-composites for real-time NIR applications," Doctoral thesis - Universität Köln, 2012.

[42] "SOITEC.com" [Online] Available: http://www.soitec.com.

[43] C. Koos, P. Vorreau, T. Vallaitis, P. Dumon, W. Bogaerts, R. Baets, B. Esembeson, I. Biaggio, T. Michinobu, F. Diederich, W. Freude, and J. Leuthold, "All-optical high-speed signal processing with silicon-organic hybrid slot waveguidesx," *Nat. Photonics*, vol. 3, no. 4, pp. 216–219, Apr. 2009.

[44] C. Koos, L. Jacome, C. Poulton, J. Leuthold, and W. Freude, "Nonlinear silicon-on-insulator waveguides for all-optical signal processing," *Opt. Express*, vol. 15, no. 10, pp. 5976–5990, 2007.

[45] D. Korn, M. Lauermann, P. Appel, L. Alloatti, R. Palmer, W. Freude, J. Leuthold, and C. Koos, "First Silicon-Organic Hybrid Laser at Telecommunication Wavelength," in *CLEO: Science and Innovations*, paper CTu2I.1, 2012.

[46] T. W. Baehr-Jones and M. J. Hochberg, "Polymer Silicon Hybrid Systems: A Platform for Practical Nonlinear Optics," *J. Phys. Chem. C*, vol. 112, no. 21, pp. 8085–8090, May 2008.

[47] J. Witzens, T. Baehr-Jones, and M. Hochberg, "Design of transmission line driven slot waveguide Mach-Zehnder interferometers and application to analog optical links," *Opt. Express*, vol. 18, no. 16, pp. 16902 16928, 2010.

[48] D. Taillaert, F. Van Laere, M. Ayre, W. Bogaerts, D. Van Thourhout, P. Bienstman, and R. Baets, "Grating Couplers for Coupling between Optical Fibers and Nanophotonic Waveguides," *Jpn. J. Appl. Phys.*, vol. 45, no. 8A, pp. 6071–6077, 2006.

[49] G. Roelkens, D. Van Thourhout, and R. Baets, "High efficiency Silicon-on-Insulator grating coupler based on a poly-Silicon overlay," *Opt. Express*, vol. 14, no. 24, pp. 11622–11630, Nov. 2006.

[50] C. Kopp, E. Augendre, R. Orobtchouk, O. Lemonnier, and J.-M. Fedeli, "Enhanced Fiber Grating Coupler Integrated by Wafer-to-Wafer Bonding," *J. Light. Technol.*, vol. 29, no. 12, pp. 1847–1851, Jun. 2011.

[51] F. Van Laere, G. Roelkens, M. Ayre, J. Schrauwen, D. Taillaert, D. Van Thourhout, T. F. Krauss, and R. Baets, "Compact and Highly Efficient Grating Couplers Between Optical Fiber and Nanophotonic Waveguides," *Light. Technol. J.*, vol. 25, no. 1, pp. 151–156, 2007.

[52] W. S. Zaoui, M. F. Rosa, W. Vogel, M. Berroth, J. Butschke, and F. Letzkus, "Cost-effective CMOS-compatible grating couplers with backside metal mirror and 69% coupling efficiency," *Opt. Express*, vol. 20, no. 26, pp. B238–B243, 2012.

[53] M. Gnan, S. Thoms, D. S. Macintyre, R. M. De La Rue, and M. Sorel, "Fabrication of low-loss photonic wires in silicon-on-insulator using hydrogen silsesquioxane electron-beam resist," *Electron. Lett.*, vol. 44, no. 2, p. 115, 2008.

[54] M. Borselli, T. J. Johnson, and O. Painter, "Measuring the role of surface chemistry in silicon microphotonics," *Appl. Phys. Lett.*, vol. 88, no. 13, p. 131114, 2006.

[55] A. Säynätjoki, L. Karvonen, T. Alasaarela, D. Korn, L. Alloatti, A. Tervonen, R. Palmer, C. Koos, J. Leuthold, W. Freude, and S. Honkanen, "Loss reduction of silicon slot waveguides with ALD grown thin films," *Proc. SPIE*, vol. 8266. p. 82660D–82660D–7, 2012.

[56] H. Seidel, "Anisotropic Etching of Crystalline Silicon in Alkaline Solutions," *J. Electrochem. Soc.*, vol. 137, no. 11, p. 3612, 1990.

[57] J. Xia, "Thermo-optic variable optical attenuator with low power consumption fabricated on silicon-on-insulator by anisotropic chemical etching," *Opt. Eng.*, vol. 43, no. 4, p. 789, Apr. 2004.

[58] W. Kern, "The Evolution of Silicon Wafer Cleaning Technology," *J. Electrochem. Soc.*, vol. 137, no. 6, pp. 1887–1892, 1990.

[59] D. Taillaert, "Grating Couplers as Interface between Optical Fibres and Nanophotonic Waveguides," Doctoral thesis - Ghent university, 2004.

[60] Y. Tang, Z. Wang, U. Westergren, and L. Wosinski, "High Efficiency Nonuniform Grating Coupler by Utilizing the Lag Effect in the Dry Etching Process," in *Optical Fiber Communication Conference (OFC 2010)*, paper OWJ6, 2010.

[61] K. A. Bates, L. Li, R. L. Roncone, and J. J. Burke, "Gaussian beams from variable groove depth grating couplers in planar waveguides," *Appl. Opt.*, vol. 32, no. 12, p. 2112, Apr. 1993.

[62] R. Waldhäusl, B. Schnabel, P. Dannberg, E.-B. Kley, A. Bräuer, and W. Karthe, "Efficient Coupling into Polymer Waveguides by Gratings," *Appl. Opt.*, vol. 36, no. 36, p. 9383, Dec. 1997.

[63] "Finisar - Active Cables" [Online] Available: http://www.finisar.com/products/active-cables/C.wire

[64] D. M. Kuchta, C. L. Schow, A. V Rylyakov, J. E. Proesel, F. E. Doany, C. Baks, B. H. Hamel-Bissell, C. Kocot, L. Graham, R. Johnson, G. Landry, E. Shaw, A. MacInnes, and J. Tatum, "A 56.1Gb/s NRZ modulated 850nm VCSEL-based optical link," in *Optical Fiber Communication Conference and Exposition and the National Fiber Optic Engineers Conference (OFC/NFOEC 2013)*, 2013.

[65] J. Guenter, B. Hawkins, R. Hawthorne, and G. Landry, "Reliability of VCSELs for >25Gb/s," in *Optical Fiber Communication Conference (OFC 2014),* 2014.

[66] D. Miller, "Energy consumption in optical modulators for interconnects," *Opt. Express*, vol. 20, no. March, pp. A293–308, Mar. 2012.

[67] P. Dong, C. Xie, L. Chen, N. K. Fontaine, and Y. Chen, "Experimental demonstration of microring quadrature phase-shift keying modulators," *Opt. Lett.*, vol. 37, no. 7, pp. 1178–1180, Apr. 2012.

[68] P. A. Sullivan and L. R. Dalton, "Theory-inspired development of organic electro-optic materials.," *Acc. Chem. Res.*, vol. 43, no. 1, Jan. 2010.

[69] D. L. Elder, S. J. Benight, J. Song, B. H. Robinson, and L. R. Dalton, "Matrix-Assisted Poling of Monolithic Bridge-Disubstituted Organic NLO Chromophores," *Chem. Mater.*, vol. 26, no. 2, pp. 872–874, Jan. 2014.

[70] T. Baehr-Jones, B. Penkov, J. Huang, P. Sullivan, J. Davies, J. Takayesu, J. Luo, T.-D. Kim, L. Dalton, A. Jen, M. Hochberg, and A. Scherer, "Nonlinear polymer-clad silicon slot waveguide modulator with a half wave voltage of 0.25 V," *Appl. Phys. Lett.*, vol. 92, no. 16, p. 163303, 2008.

[71] R. Palmer, S. Koeber, M. Woessner, D. L. Elder, W. Heni, D. Korn, H. Yu, M. Lauermann, W. Bogaerts, L. R. Dalton, W. Freude, J. Leuthold, and C. Koos, "High-Speed Silicon-Organic Hybrid (SOH) Modulators with 230 pm/V Electro-Optic Coefficient Using Advanced Materials," in *Optical Fiber Communication Conference (OFC 2014),* paper M3G.4, 2014.

[72] R. Palmer, S. Koeber, W. Heni, D. L. Elder, D. Korn, H. Yu, L. Alloatti, S. Koenig, P. C. Schindler, W. Bogaerts, M. Pantouvaki, G. Lepage, P. Verheyen, J. Van Campenhout, P. Absil, R. Baets, L. R. Dalton, W. Freude, J. Leuthold, and C. Koos, "High-Speed Silicon-Organic Hybrid (SOH) Modulator with 1.6 fJ/bit and 180 pm/V In-Device Nonlinearity," in *39th European Conference and Exhibition on Optical Communication (ECOC 2013)*, 2013.

[73] C. Weimann, S. Wolf, D. Korn, R. Palmer, S. Koeber, R. Schmogrow, P. C. Schindler, L. Alloatti, A. Ludwig, W. Heni, D. Bekele, D. L. Elder, H. Yu, W. Bogaerts, L. R. Dalton, W. Freude, J. Leuthold, and C. Koos, "Silicon-organic hybrid (SOH) frequency comb source for data transmission at 784 Gbit/s," in *39th European Conference and Exhibition on Optical Communication (ECOC 2013)*, 2013.

[74] C. Weimann, P. C. Schindler, R. Palmer, S. Wolf, D. Bekele, D. Korn, J. Pfeifle, S. Koeber, R. Schmogrow, L. Alloatti, D. Elder, H. Yu, W. Bogaerts, L. R. Dalton, W. Freude, J. Leuthold, and C. Koos, "Silicon-organic hybrid (SOH) frequency comb sources for terabit/s data transmission," *Opt. Express*, vol. 22, no. 3, pp. 3629–3637, Feb. 2014.

[75] C. Weimann, P. Schindler, D. Bekele, R. Palmer, D. Korn, J. Pfeifle, S. Koeber, R. Schmogrow, L. Alloatti, D. L. Elder, H. Yu, W. Bogaerts, L. R. Dalton, W. Freude, J. Leuthold, and C. Koos, "Data Transmission at Terabit/s Data Rates Using Silicon-Organic Hybrid (SOH) Frequency Combs," in *Optical Fiber Communication Conference (OFC 2014)*, paper Th4I.2, 2014.

[76] M. Lauermann, C. Weimann, A. Knopf, D. L. Elder, W. Heni, R. Palmer, D. Korn, P. Schindler, S. Koeber, L. Alloatti, H. Yu, W. Bogaerts, L. R. Dalton, C. Rembe, J. Leuthold, W. Freude, and C. Koos, "Integrated Silicon-Organic Hybrid (SOH) Frequency Shifter," in *Optical Fiber Communication Conference (OFC 2014)*, paper Tu2A.1, 2014.

[77] J. Takayesu, M. Hochberg, T. Baehr-Jones, E. Chan, P. Sullivan, J. Davies, L. Dalton, A. Scherer, and W. Krug, "A Hybrid Electrooptic Microring Resonator-Based 1x4x1 ROADM for Wafer Scale Optical Interconnects," *J. Light. Technol.*, vol. 27, no. 4, pp. 440–448, Feb. 2009.

[78] Y. Liao, C. A. Anderson, P. A. Sullivan, A. J. P. Akelaitis, B. H. Robinson, and L. R. Dalton, "Electro-Optical Properties of Polymers Containing Alternating Nonlinear Optical Chromophores and Bulky Spacers," *Chem. Mater.*, vol. 18, no. 4, pp. 1062–1067, 2006.

[79] J. Luo, X.-H. Zhou, and A. K.-Y. Jen, "Rational molecular design and supramolecular assembly of highly efficient organic electro-optic materials," *J. Mater. Chem.*, vol. 19, no. 40, pp. 7410–7424, 2009.

[80] X. Wang, C.-Y. Lin, S. Chakravarty, J. Luo, A. K.-Y. Jen, and R. T. Chen, "Effective in-device r33 of 735 pm/V on electro-optic polymer infiltrated silicon photonic crystal slot waveguides," *Opt. Lett.*, vol. 36, no. 6, pp. 882–884, 2011.

[81] M. Gould, T. Baehr-Jones, R. Ding, S. Huang, J. Luo, A. K.-Y. Jen, J.-M. Fedeli, M. Fournier, and M. Hochberg, "Silicon-polymer hybrid slot waveguide ring-resonator modulator.," *Opt. Express*, vol. 19, no. 5, pp. 3952–3961, 2011.

[82] D. Jin, H. Chen, A. Barklund, J. Mallari, and G. Yu, "EO polymer modulators reliability study," *Proc. SPIE*, 2010.

[83] M. Bachmann, P. A. Besse, and H. Melchior, "General self-imaging properties in N x N multimode interference couplers including phase relations," *Appl. Opt.*, vol. 33, no. 18, pp. 3905–3911, Jun. 1994.

[84] J. Leuthold, W. Freude, J.-M. Brosi, R. Baets, P. Dumon, I. Biaggio, M. L. Scimeca, F. Diederich, B. Frank, and C. Koos, "Silicon Organic Hybrid Technology—A Platform for Practical Nonlinear Optics," *Proc. IEEE*, vol. 97, no. 7, pp. 1304–1316, Jul. 2009.

[85] A. E. Vasdekis, S. A. Moore, A. Ruseckas, T. F. Krauss, I. D. W. Samuel, and G. A. Turnbull, "Silicon based organic semiconductor laser," *Appl. Phys. Lett.*, vol. 91, no. 5, Jul. 2007.

[86] M. Hochberg, T. Baehr-Jones, G. Wang, J. Huang, P. Sullivan, L. Dalton, and A. Scherer, "Towards a millivolt optical modulator with nano-slot waveguides," *Opt. Express*, vol. 15, no. 13, pp. 8401–8410, Jun. 2007.

[87] L. Alloatti, D. Korn, C. Weimann, C. Koos, W. Freude, and J. Leuthold, "Second-order nonlinear silicon-organic hybrid waveguides," *Opt. Express*, vol. 20, no. 18, pp. 20506–20515, Aug. 2012.

[88] J. Pfeifle, L. Alloatti, W. Freude, J. Leuthold, and C. Koos, "Silicon-organic hybrid phase shifter based on a slot waveguide with a liquid-crystal cladding," *Opt. Express*, vol. 20, no. 14, p. 15359, 2012.

[89] T. Alasaarela, D. Korn, L. Alloatti, A. Säynätjoki, A. Tervonen, R. Palmer, J. Leuthold, W. Freude, and S. Honkanen, "Reduced propagation loss in silicon strip and slot waveguides coated by atomic layer deposition," *Opt. Express*, vol. 19, no. 12, pp. 11529–11538, Jun. 2011.

[90] Z. Wang, N. Zhu, Y. Tang, L. Wosinski, D. Dai, and S. He, "Ultracompact low-loss coupler between strip and slot waveguides," *Opt. Lett.*, vol. 34, no. 10, pp. 1498–1500, 2009.

[91] Y. Liu, T. Baehr-Jones, J. Li, A. Pomerene, and M. Hochberg, "Efficient Strip to Strip-Loaded Slot Mode Converter in Silicon-on-Insulator," *IEEE Photonics Technol. Lett.*, vol. 23, no. 20, 2011.

[92] A. Milton and W. K. Burns, "Mode coupling in optical waveguide horns," *Quantum Electron. IEEE J.*, vol. 13, no. 10, pp. 828–835, 1977.

[93] D. M. Pozar, *Microwave Engineering*, 2nd ed. John Wiley & Sons, Inc., 1998.

[94] D. Marcuse, "Radiation Losses of Tapered Dielectric Slab Waveguides," *Bell Syst. Tech. J.*, vol. 49, no. 2, pp. 273–290, Feb. 1970.

[95] D. F. G. Gallagher and T. P. Felici, "Eigenmode expansion methods for simulation of optical propagation in photonics: pros and cons," *Proc. SPIE*, vol. 4987. pp. 69–82, 2003.

[96] G. Ren, S. Chen, Y. Cheng, and Y. Zhai, "Study on inverse taper based mode transformer for low loss coupling between silicon wire waveguide and lensed fiber," *Opt. Commun.*, vol. 284, no. 19, pp. 4782–4788, Sep. 2011.

[97] L. R. Dalton, P. a Sullivan, and D. H. Bale, "Electric field poled organic electro-optic materials: state of the art and future prospects.," *Chem. Rev.*, vol. 110, no. 1, pp. 25–55, Jan. 2010.

[98] Y. V. Pereverzev, K. N. Gunnerson, O. V. Prezhdo, P. A. Sullivan, Y. Liao, B. C. Olbricht, A. J. P. Akelaitis, A. K.-Y. Jen, and L. R. Dalton, "Guest–Host Cooperativity in Organic Materials Greatly Enhances the Nonlinear Optical Response," *J. Phys. Chem. C*, vol. 112, no. 11, pp. 4355–4363, Mar. 2008.

[99] X. Wang, C.-Y. Lin, S. Chakravarty, J. Luo, A. K.-Y. Jen, and R. T. Chen, "Effective in-device r33 of 735 pm/V on electro-optic polymer infiltrated silicon photonic crystal slot waveguides.," *Opt. Lett.*, vol. 36, no. 6, pp. 882–884, 2011.

[100] Y. Enami, C. T. Derose, D. Mathine, C. Loychik, C. Greenlee, R. Norwood, T. D. Kim, J. Luo, Y. Tian, a. K.-Y. Jen, and N. Peyghambarian, "Hybrid polymer/sol–gel waveguide modulators with exceptionally large electro–optic coefficients," *Nat. Photonics*, vol. 1, no. 3, pp. 180–185, Mar. 2007.

[101] L. Chen, K. Preston, S. Manipatruni, and M. Lipson, "Integrated GHz silicon photonic interconnect with micrometer-scale modulators and detectors," *Opt. Express*, vol. 17, no. 17, pp. 15248–15256, Aug. 2009.

[102] S. Manipatruni, Q. Xu, B. Schmidt, J. Shakya, and M. Lipson, "High Speed Carrier Injection 18 Gb/s Silicon Micro-ring Electro-optic Modulator," in *Lasers and Electro-Optics Society, 2007. LEOS 2007. The 20th Annual Meeting of the IEEE*, pp. 537–538, 2007.

[103] P. Dong, S. Liao, D. Feng, H. Liang, D. Zheng, R. Shafiiha, C.-C. Kung, W. Qian, G. Li, X. Zheng, A. V Krishnamoorthy, and M. Asghari, "Low Vpp, ultralow-energy, compact, high-speed silicon electro-optic modulator," *Opt. Express*, vol. 17, no. 25, pp. 22484–22490, 2009.

[104] L. Liao, A. Liu, J. Basak, H. Nguyen, M. Paniccia, D. Rubin, Y. Chetrit, R. Cohen, and N. Izhaky, "40 Gbit/s silicon optical modulator for highspeed applications," *Electron. Lett.*, vol. 43, no. 22, 2007.

[105] F. Y. Gardes, D. J. Thomson, N. G. Emerson, and G. T. Reed, "40 Gb/s silicon photonics modulator for TE and TM polarisations," *Opt. Express*, vol. 19, no. 12, pp. 11804–11814, Jun. 2011.

[106] P. Dong, L. Chen, C. Xie, L. L. Buhl, and Y.-K. Chen, "50-Gb/s silicon quadrature phase-shift keying modulator," *Opt. Express*, vol. 20, no. 19, pp. 21181–21186, Sep. 2012.

[107] D. J. Thomson, F. Y. Gardes, Y. Hu, G. Mashanovich, M. Fournier, P. Grosse, J.-M. Fedeli, and G. T. Reed, "High contrast 40Gbit/s optical modulation in silicon," *Opt. Express*, vol. 19, no. 12, pp. 11507–11516, Jun. 2011.

[108] K. Ogawa, K. Goi, Y. T. Tan, T.-Y. Liow, X. Tu, Q. Fang, G.-Q. Lo, and D.-L. Kwong, "Silicon Mach-Zehnder modulator of extinction ratio beyond 10 dB at 10.0-12.5 Gbps," *Opt. Express*, vol. 19, no. 26, pp. B26–B31, 2011.

[109] R. Palmer, L. Alloatti, D. Korn, W. Heni, P. C. Schindler, J. Bolten, M. Karl, M. Waldow, T. Wahlbrink, W. Freude, C. Koos, and J. Leuthold, "Low-Loss Silicon Strip-to-Slot Mode Converters," *IEEE Photonics J.*, vol. 5, no. 1, p. 2200409, Feb. 2013.

[110] R. Schmogrow, D. Hillerkuss, M. Dreschmann, M. Huebner, M. Winter, J. Meyer, B. Nebendahl, C. Koos, J. Becker, W. Freude, and J. Leuthold, "Real-Time Software-Defined Multiformat Transmitter Generating 64QAM at 28 GBd," *IEEE PHOTONICS Technol. Lett.*, vol. 22, no. 21, pp. 1601–1603, Nov. 2010.

[111] R. Schmogrow, B. Nebendahl, M. Winter, A. Josten, D. Hillerkuss, S. Koenig, J. Meyer, M. Dreschmann, M. Huebner, C. Koos, J. Becker, W. Freude, and J. Leuthold, "Error Vector Magnitude as a Performance Measure for Advanced Modulation Formats," *IEEE Phot. Technol. Lett.*, vol. 24, no. 1, pp. 61–63, Jan. 2012.

[112] R. Schmogrow, B. Nebendahl, M. Winter, A. Josten, D. Hillerkuss, S. Koenig, J. Meyer, M. Dreschmann, M. Huebner, C. Koos, J. Becker, W. Freude, and J. Leuthold, "Corrections to "Error Vector Magnitude as a Performance Measure for Advanced Modulation Formats," *IEEE Phot. Technol. Lett.*, vol. 24, no. 23, p. 2198, Dec. 2012.

[113] T. Mizuochi, "Recent progress in forward error correction and its interplay with transmission impairments," *IEEE J. Sel. Top. Quantum Electron.*, vol. 12, no. 4, pp. 544–554, 2006.

[114] G. Bosco, V. Curri, A. Carena, P. Poggiolini, and F. Forghieri, "On the Performance of Nyquist-WDM Terabit Superchannels Based on PM-BPSK, PM-QPSK, PM-8QAM or PM-16QAM Subcarriers," *Light. Technol. J.*, vol. 29, no. 1, pp. 53–61, 2011.

[115] R. Schmogrow, M. Winter, M. Meyer, D. Hillerkuss, S. Wolf, B. Baeuerle, A. Ludwig, B. Nebendahl, S. Ben-Ezra, J. Meyer, M. Dreschmann, M. Huebner, J. Becker, C. Koos, W. Freude, and J. Leuthold, "Real-time Nyquist pulse generation beyond 100 Gbit/s and its relation to OFDM," *Opt. Express*, vol. 20, no. 1, pp. 317–337, Jan. 2012.

[116] D. Hillerkuss, R. Schmogrow, M. Meyer, S. Wolf, M. Jordan, P. Kleinow, N. Lindenmann, P. Schindler, A. Melikyan, X. Yang, S. Ben-Ezra, B. Nebendahl, M. Dreschmann, J. Meyer, F. Parmigiani, P. Petropoulos, B. Resan, A. Oehler, K. Weingarten, L. Altenhain, T. Ellermeyer, M. Moeller, M. Huebner, J. Becker, C. Koos, W. Freude, and J. Leuthold, "Single-Laser 32.5 Tbit/s Nyquist WDM Transmission," *J. Opt. Commun. Netw.*, vol. 4, no. 10, pp. 715–723, 2012.

[117] R. Schmogrow, M. Meyer, S. P., A. Josten, S. Ben-Ezra, C. Koos, W. Freude, and J. Leuthold, "252 Gbit/s Real-Time Nyquist Pulse Generation by Reducing the Oversampling Factor to 1.33," in *Optical Fiber Communication Conference (OFC 2013), paper OTu2I.1*, 2013.

[118] L. Liao, A. Liu, D. Rubin, J. Basak, Y. Chetrit, H. Nguyen, R. Cohen, N. Izhaky, and M. Paniccia, "40 Gbit/s silicon optical modulator for high- speed applications," *Electron. Lett.*, vol. 43, no. 22, p. 1196, 2007.

[119] D. J. Thomson, F. Y. Gardes, J.-M. Fedeli, S. Zlatanovic, Y. Hu, B. P. P. Kuo, E. Myslivets, N. Alic, S. Radic, G. Z. Mashanovich, and G. T. Reed, "50-Gb/s Silicon Optical Modulator," *Phot. Technol. Lett. IEEE*, vol. 24, no. 4, pp. 234–236, 2012.

[120] J. Fujikata, J. Ushida, T. Nakamura, Y. Ming-Bin, Z. ShiYang, D. Liang, P. L. Guo-Qiang, and D.-L. Kwong, "25 GHz Operation of Silicon Optical Modulator with Projection MOS Structure," *Opt. Fiber Commun. Conf. (OFC 2010)*, paper OMI3, 2010.

[121] C. A. Barrios, V. R. Almeida, R. Panepucci, and M. Lipson, "Electrooptic Modulation of Silicon-on-Insulator Submicrometer-Size Waveguide Devices," *J. Light. Technol.*, vol. 21, no. 10, p. 2332, 2003.

[122] J. C. Rosenberg, W. M. Green, S. Assefa, T. Barwicz, M. Yang, S. M. Shank, and Y. A. Vlasov, "Low-Power 30 Gbps Silicon Microring Modulator," in *CLEO:2011 - Laser Applications to Photonic Applications*, paper PDPB9, 2011.

[123] T. Vallaitis, S. Bogatscher, L. Alloatti, P. Dumon, R. Baets, M. L. Scimeca, I. Biaggio, F. Diederich, C. Koos, W. Freude, and J. Leuthold, "Optical properties of highly nonlinear silicon-organic hybrid (SOH) waveguide geometries," *Opt. Express*, vol. 17, no. 20, pp. 17357–17368, Sep. 2009.

[124] C. M. Osburn and D. W. Ormond, "Dielectric Breakdown in Silicon Dioxide Films on Silicon," *J. Electrochem. Soc.*, vol. 119(5), pp. 597–603, 1972.

[125] D. A. B. Miller, "Energy consumption in optical modulators for interconnects," *Opt. Express*, vol. 20, no. S2, pp. A293–A308, 2012.

[126] W. M. Green, M. J. Rooks, L. Sekaric, and Y. A. Vlasov, "Ultra-compact, low RF power, 10 Gb/s silicon Mach-Zehnder modulator," *Opt. Express*, vol. 15, no. 25, p. 17106, 2007.

[127] T. Baba, S. Akiyama, M. Imai, T. Akagawa, M. Takahashi, N. Hirayama, H. Takahashi, Y. Noguchi, H. Okayama, T. Horikawa, and T. Usuki, "25-Gbps operation of silicon p-i-n Mach-Zehnder optical modulator with 100-µm-long phase shifter," in *CLEO: Science and Innovations*, paper CF2L.3, 2012.

[128] V. R. Almeida, Q. Xu, C. A. Barrios, and M. Lipson, "Guiding and confining light in void nanostructure," *Opt. Lett.*, vol. 29, no. 11, p. 1209, 2004.

[129] J. Leuthold, W. Freude, J.-M. Brosi, R. Baets, P. Dumon, I. Biaggio, M. L. Scimeca, F. Diederich, B. Frank, and C. Koos, "Silicon organic hybrid technology - A platform for practical nonlinear optics," *Proc. IEEE*, vol. 97, no. 7, pp. 1304–1316, Jul. 2009.

[130] J.-M. Brosi, C. Koos, L. C. Andreani, M. Waldow, J. Leuthold, and W. Freude, "High-speed low-voltage electro-optic modulator with a polymer-infiltrated silicon photonic crystal waveguide," *Opt. Express*, vol. 16, no. 6, p. 4177, Mar. 2008.

[131] M. Hochberg, T. Baehr-Jones, G. Wang, J. Huang, P. Sullivan, L. Dalton, and A. Scherer, "Towards a millivolt optical modulator with nano-slot waveguides," *Opt. Express*, vol. 15, no. 13, p. 8401, Jun. 2007.

[132] J. H. Wülbern, S. Prorok, J. Hampe, A. Petrov, M. Eich, J. Luo, A. K.-Y. Jen, M. Jenett, and A. Jacob, "40 GHz electro-optic modulation in hybrid silicon-organic slotted photonic crystal waveguides.," *Opt. Lett.*, vol. 35, no. 16, pp. 2753–5, Aug. 2010.

[133] R. Ding, T. Baehr-Jones, Y. Liu, R. Bojko, J. Witzens, S. Huang, J. Luo, S. Benight, P. Sullivan, J.-M. Fedeli, M. Fournier, L. Dalton, a Jen, and M. Hochberg, "Demonstration of a low V pi L modulator with GHz bandwidth based on electro-optic polymer-clad silicon slot waveguides.," *Opt. Express*, vol. 18, no. 15, pp. 15618–23, Jul. 2010.

[134] R. Palmer, L. Alloatti, D. Korn, P. C. Schindler, M. Baier, J. Bolten, T. Wahlbrink, M. Waldow, R. Dinu, W. Freude, C. Koos, and J. Leuthold, "Low Power Mach-Zehnder Modulator in Silicon-Organic Hybrid Technology," *IEEE Photonics Technol. Lett.*, vol. 25, no. 13, pp. 1226–1229, Jul. 2013.

[135] J. J. Wolff and R. Wortmann, "Organic Materials for Second-Order Non-Linear Optics," in *Advances in Physical Organic Chemistry*, vol. 32, Elsevier, 1999, pp. 121–217.

[136] L. R. Dalton, S. J. Benight, L. E. Johnson, D. B. Knorr, I. Kosilkin, B. E. Eichinger, B. H. Robinson, A. K.-Y. Jen, and R. M. Overney, "Systematic Nanoengineering of Soft Matter Organic Electro-optic Materials," *Chem. Mater.*, vol. 23, no. 3, pp. 430–445, Feb. 2011.

[137] L. R. Dalton, P. A. Sullivan, D. Bale, B. Olbricht, J. Davies, I. Kosilkin, B. H. Robinson, B. E. Eichinger, and A. K.-Y. Jen, "Organic electro-optic materials," in *Organic Thin Films for Photonic Applications*, vol. 1039, pp. 13–33, 2010.

[138] T. Baehr-Jones, M. Hochberg, G. Wang, R. Lawson, Y. Liao, P. Sullivan, L. Dalton, a Jen, and a Scherer, "Optical modulation and detection in slotted Silicon waveguides.," *Opt. Express*, vol. 13, no. 14, pp. 5216–26, Jul. 2005.

[139] S. Huang, J. Luo, H.-L. Yip, A. Ayazi, X.-H. Zhou, M. Gould, A. Chen, T. Baehr-Jones, M. Hochberg, and A. K.-Y. Jen, "Efficient poling of electro-optic polymers in thin films and silicon slot waveguides by detachable pyroelectric crystals," *Adv. Mater.*, vol. 24, no. 10, pp. OP42–7, Mar. 2012.

[140] S. J. Benight, L. E. Johnson, R. Barnes, B. C. Olbricht, D. H. Bale, P. J. Reid, B. E. Eichinger, L. R. Dalton, P. a Sullivan, and B. H. Robinson, "Reduced dimensionality in organic electro-optic

materials: theory and defined order.," *J. Phys. Chem. B*, vol. 114, no. 37, pp. 11949–56, Sep. 2010.

[141] R. Palmer, S. Koeber, D. Elder, M. Woessner, W. Heni, D. Korn, M. Lauermann, W. Bogaerts, L. Dalton, W. Freude, J. Leuthold, and C. Koos, "High-Speed, Low Drive-Voltage Silicon-Organic Hybrid Modulator based on a Binary-Chromophore Electro-Optic Material," *J. Light. Technol.*, 2014.

[142] P. Dong, L. Chen, and Y. Chen, "High-speed low-voltage single-drive push-pull silicon Mach-Zehnder modulators," *Opt. Express*, vol. 20, no. 6, pp. 6163–6169, 2012.

[143] J. Leuthold, C. Koos, W. Freude, L. Alloatti, R. Palmer, D. Korn, J. Pfeifle, M. Lauermann, R. Dinu, S. Wehrli, M. Jazbinsek, P. Gunter, M. Waldow, T. Wahlbrink, J. Bolten, H. Kurz, M. Fournier, J.-M. Fedeli, H. Yu, and W. Bogaerts, "Silicon-Organic Hybrid Electro-Optical Devices," *IEEE J. Sel. Top. Quantum Electron.*, vol. 19, no. 6, pp. 3401413–3401413, Nov. 2013.

[144] F. Chang, K. Onohara, and T. Mizuochi, "Forward error correction for 100 G transport networks," *IEEE Commun. Mag.*, vol. 48, no. 3, pp. S48–S55, Mar. 2010.

[145] J. Luo, S. Huang, Z. Shi, B. M. Polishak, X.-H. Zhou, and A. K. Jen, "Tailored Organic Electro-optic Materials and Their Hybrid Systems for Device Applications," *Chem. Mater.*, vol. 23, no. 3, pp. 544–553, Feb. 2011.

[146] Z. Shi, W. Liang, J. Luo, S. Huang, B. M. Polishak, X. Li, T. R. Younkin, B. a. Block, and A. K.-Y. Jen, "Tuning the Kinetics and Energetics of Diels−Alder Cycloaddition Reactions to Improve Poling Efficiency and Thermal Stability of High-Temperature Cross-Linked Electro-Optic Polymers," *Chem. Mater.*, vol. 22, no. 19, pp. 5601–5608, Oct. 2010.

[147] W. Huang, Z. Jin, Z. Shi, J. J. Intemann, M. Li, J. Luo, and A. K.-Y. Jen, "Spontaneous thermal crosslinking of a sydnone-containing side-chain polymer with maleimides through a convergent [3 + 2] dual cycloaddition/cycloreversion process for electro-optics," *Polym. Chem.*, vol. 4, no. 24, p. 5760, 2013.

[148] Z. Shi, J. Luo, S. Huang, B. M. Polishak, X.-H. Zhou, S. Liff, T. R. Younkin, B. A. Block, and A. K.-Y. Jen, "Achieving excellent electro-optic activity and thermal stability in poled polymers through an

expeditious crosslinking process," *J. Mater. Chem.*, vol. 22, no. 3, p. 951, 2012.

[149] R. A. Soref and B. R. Bennett, "Electrooptical effects in silicon," *Quantum Electron. IEEE J.*, vol. 23, no. 1, pp. 123–129, Jan. 1987.

[150] R. Ding, T. Baehr-Jones, W.-J. Kim, X. Xiong, R. Bojko, J.-M. Fedeli, M. Fournier, and M. Hochberg, "Low-loss strip-loaded slot waveguides in Silicon-on-Insulator," *Opt. Express*, vol. 18, no. 24, pp. 25061–25067, Nov. 2010.

[151] R. Ding, T. Baehr-Jones, W.-J. Kim, B. Boyko, R. Bojko, A. Spott, A. Pomerene, C. Hill, W. Reinhardt, and M. Hochberg, "Low-loss asymmetric strip-loaded slot waveguides in silicon-on-insulator," *Appl. Phys. Lett.*, vol. 98, no. 23, 2011.

[152] D. F. Bishop, S. Million, T. M. Nguyen, and M. K. Simon, "Power spectrum of unbalanced NRZ and biphase signals in the presence of data asymmetry," *Electromagn. Compat. IEEE Trans.*, vol. 40, no. 1, pp. 55–61, 1998.

[153] D. Marcuse, *Light Transmission Optics*. Van Nostrand Reinhold Company, 1972.

[154] L. B. Soldano and E. C. M. Pennings, "Optical Multi-Mode Interference Devices Based on Self-Imaging : Principles and Applications," *J. Light. Technol.*, vol. 13, no. 4, pp. 615–627, 1995.

[155] C. Koos, "Nanophotonic Devices for Linear and Nonlinear Optical Signal Processing," Doctoral thesis - Universität Karlsruhe, 2007.

[156] D. Marcuse, *Theory of dielectric optical waveguides*. Academic Press, 1974.

[157] T. Vallaitis, "Ultrafast Nonlinear Silicon Waveguides and Quantum Dot Semiconductor Optical Amplifiers," Doctoral thesis - Karlsruhe Institute of Technology, 2011.

[158] G. Grau and W. Freude, *Optische Nachrichtentechnik*, 3rd ed. Springer, 2005.

[159] "Photonics Wiki" [Online] Available: http://photonicswiki.org/index.php?title=File:Poling_centric.JPG

Appendix I Glossary

I.1 Crystallographic Notation

(ijk)-planes	Multitude of parallel crystal planes with normal vector $[ijk]$
$\{ijk\}$-planes	Multitude of crystal planes that are equivalent to the planes (ijk) by crystal symmetry
$[ijk]$	Crystal-vector with coordinates i,j,k
$<ijk>$	Multitude of crystal-vectors that are equivalent to vector $[ijk]$ by crystal symmetry

I.2 List of Symbols

Greek Symbols

$\left\langle \cos^3\theta \right\rangle$	Average acentric order parameter
$\underline{\alpha}$	First-order polarizability
α_c	Power attenuation coefficient of the optical carrier in m^{-1}
α_l	Power leakage parameter of a grating in m^{-1}
α_m	Power attenuation coefficient of a microwave in m^{-1}
β	Propagation constant of an electro-magnetic wave
$\underline{\beta}$	Second-order polarizability/first hyperpolarizability
γ	Complex propagation constant of an EM wave
$\underline{\gamma}$	Third-order polarizability/second hyperpolarizability
Γ	Field interaction factor
δ_{ij}	Kronecker-delta
δ_s	Skin depth
Δ	Relative refractive index difference
ε_0	Electric permittivity of vacuum
ε_r	Relative electric permittivity of a material
$\Delta\varepsilon_1$	Change of relative permittivity
$\vec{\mu}$	Electric dipole moment

μ_0	Magnetic permeability of free space
μ_r	Relative magnetic permeability of a material
η	Coupling efficiency between fiber and SOI chip
ϑ	Sidewall angle of a trapezoidal waveguide
ϑ_{uu}	Autocorrelation function
θ	Angle between input electric field polarization and molecular dipole axis
$\Theta_0, \Theta_c, \Theta_r$	Power spectral density at the source, at the device capacitor, and at the series resistor
$\kappa^{(P)}$	Power coupling coefficient
$\kappa_{\mu\nu}$	Mode coupling coefficient
λ_c	Optical wavelength
Λ	Grating period
ϱ	Coupling angle between fiber and chip with respect to the chip surface normal
ρ	Electrical resistivity
τ	RC time constant
φ, $\delta\varphi$	Phase and induced phase change due to Pockels effect
$\varphi_1, \varphi_2, \varphi_{bias}$	Phases after phase shifter 1, 2 and after the bias phase shifter in a Mach-Zehnder modulator
$\Delta\varphi$	Difference of phases $\Delta\varphi = \varphi_2 - \varphi_1 - \varphi_{bias}$
ω_c	Angular frequency of the optical carrier
ω_m	Angular frequency of the microwave
$\underline{\chi}^{(n)}$	Influence function, tensor of rank $n+1$
$\underline{\tilde{\chi}}^{(n)}$	Optical susceptibility of nth-order

Latin Symbols

B^2	Power per length that is radiated away from a grating
c	Vacuum speed of light
C_{slot}	Capacitance of a striploaded slot waveguide
C_d	Capacitance of a single-drive MZM device

D	Directionality of a grating coupler
$\vec{D}$	Displacement field vector
e	Elementary charge
$\vec{E}$	Electric field vector
$\vec{\mathcal{E}}_\mu$	Transverse electric mode field of mode μ
f_c	Optical frequency
f_m	Microwave frequency
$f_{3dB,EO}$	Electro-optic 3 dB cut-off frequency
$f_{6dB,EOE}$	Electro-optic-electric 6 dB cut-off frequency
g	Lorentz-Onsager local field factor
G	Gaussian power distribution per length
h_{dev}	Height of the device layer of a silicon-on-insulator wafer
h_{groove}	Etch depth of the grating grooves
h_{load}	Height of the stripload of a striploaded slot waveguide
$\vec{H}$	Magnetic field vector
H_c, H_r	Amplitude transfer functions
$\vec{\mathcal{H}}_\mu$	Transverse magnetic mode field of mode μ
δi_{PD}	AC current of a photo diode
δI	Magnitude of optical intensity modulation
$\vec{k}$	Wave vector
K	Linear momentum of a lattice of period Λ
L	Modulator length
L_{eff}	Effective modulator length
n, n_i	Real and imaginary part of the refractive index
n_{eff}	Effective refractive index
n_g^m, n_g^o	Group indices of microwave signal and optical signal
n_{slot}	Refractive index of the material in the slot of a silicon slot waveguide
Δn	Change of refractive index
N	Dipole number density

$\vec{p}$	Molecular polarizability
$\vec{P}$	Electric polarization vector
$\vec{P}^{(1)}$	Linear electric polarization
$\vec{P}^{(2)}$	Second-order nonlinear electric polarization
$\vec{P}_{\mathrm{NL}}$	Nonlinear electric polarization
P	Power
$\mathcal{P}_{\mu}$	Normalizing power of the optical mode μ
Q	Quadrature phase component of a signal
$\underline{r}$	Electro-optic coefficient/tensor
r	Data rate
R_{d}	Internal series resistance of a MZM device
R_{load}	Sum of stripload resistances of a phase modulator
R_{s}	Internal resistance of an RF source
R_{t}	Terminating resistance
S_{11}, S_{21}	S-parameters of a two-port structure describing reflection and transmission
T	Symbol duration of a signal
T_g	Glass-transition temperature
t	Time
$\Delta t_{\mathrm{o,m}}$	Time delay between optical signal and electrical signal at the end of a modulator
U_0	Open-circuit voltage of a source
U_{bias}	Bias voltage
U_{c}	Slot capacitor voltage. Voltage that drops across the slot of a slot waveguide
U_{drive}	AC peak-to-peak drive voltage
U_{poling}	DC poling voltage
U_{π}	π -voltage of a MZM
v	Phase velocity
$v_{\mathrm{g}}^{\mathrm{m}}, v_{\mathrm{g}}^{\mathrm{o}}$	Group velocities of microwave signal and optical signal

w_0	Gaussian beam waist, $2\,w_0$: Mode field diameter
w_{bottom}	Bottom width of a trapezoidal waveguide
w_{groove}	Width of a grating grooves
w_{rail}	Width of a rail of a slot waveguide
w_{slot}	Width of the slot of a slot waveguide
w_{strip}	Width of a strip waveguide
w_{top}	Width of a top surface of a trapezoidal waveguide
W_{bit}	Energy per bit
z	Spatial variable
Z_0	Free space wave impedance
Z_{L}	Transmission line impedance
Z_{slot}	Impedance of the slot waveguide capacitor
Z_{d}	Impedance of the MZM device

I.3 Acronyms

16QAM	16ary quadrature amplitude modulation
ASK	Amplitude shift keying
B2B	Back-to-back, characterization without transmission link
BER	Bit error ratio or bit error probability
BERT	Bit error ratio tester
BOX	Buried oxide
BPSK	Binary phase shift keying
DAC	Digital-to-analog converter
DCA	Digital communications analyzer - Agilent Oscilloscope
DSP	Digital signal processing
EBL	Electron beam lithography
ECL	External cavity laser
EDFA	Erbium doped fiber amplifier

EVM	Error vector magnitude
EVM_m	EVM normalized to the maximum field in the constellation
EO	Electro-optic
EOE	Electrical-optical-electrical
FEC	Forward-error-correction
GC	Grating coupler
HSQ	Hydrogen silsesquioxane – negative tone photo resist
IQ-modulator	Modulator for complex modulation formats
KOH	Potassium hydroxide
$LiNbO_3$	Lithium Niobate
LHS	Left-hand side
LO	Local oscillator
MEMS	Microelectromechanical systems
MMI	Multimode interference coupler
MZI	Mach-Zehnder interferometer
MZM	Mach-Zehnder modulator
NRZ	Non-return to zero
OBR	Optical backscatter reflectometry
OMA	Agilent optical modulation analyzer N4391A
OOK	On-off-keying
OSA	Optical spectrum analyzer
PDM	Polarization division multiplexing
PD	Photo diode
PIC	Photonic integrated circuit
PolMUX	Polarization multiplexing
PPG	Pseudorandom pattern generator
PRBS	Pseudorandom bit sequence
QAM	Quadrature amplitude modulation
QPSK	Quadrature phase shift keying

RF	Radio frequency
RHS	Right-hand side
rms	Root mean square
Rx	Receiver
Si	Silicon
SiO_2	Silicon dioxide
SOI	Silicon-on-insulator
SSMF	Standard singlemode fiber
TMAH	Tetramethylammonium hydroxide
Tx	Transmitter
VCSEL	Vertical cavity surface emitting laser
VOA	Variable optical attenuator
WG	Waveguide

Acknowledgements (German)

Die vorliegende Dissertation entstand während meiner Tätigkeit am Institut für Photonik und Quantenelektronik (IPQ) am Karlsruher Institut für Technologie. Die wissenschaftlichen Ergebnisse dieser Arbeit wurden im Rahmen des vom Bundesministerium für Bildung und Forschung (BMBF) geförderten Projektes MISTRAL (Förderkennzeichen 01BL0804) erarbeitet sowie im Rahmen des von der Europäischen Union (EU) geförderten FP-7 Projektes SOFI (grant 248609). Zum Abschluss bleibt mir die angenehme Aufgabe mich bei all den Personen zu bedanken die zum Gelingen dieser Arbeit beigetragen haben.

An erster Stelle bedanke ich mich bei meinen Betreuern Prof. Juerg Leuthold, Prof. Wolfgang Freude und Prof. Christian Koos für Inspiration, stetige Motivation, das entgegengebrachte Vertrauen und für die Organisation und Bereitstellung finanzieller Mittel für diese Arbeit.

Ich danke meinem Doktorvater und langjährigen Institutsleiter Prof. Juerg Leuthold, dessen visionäre Sicht und ansteckender Optimismus ein kontinuierlicher Ansporn waren. Das Gewicht, das er auf eine angemessene Vermittlung wissenschaftlicher Erkenntnisse legte, und die Vermittlung der zugehörigen Techniken waren wesentlich für den Erfolg der Veröffentlichungen. Danke auch für die vielen interessanten fachübergreifenden Gespräche.

Prof. Wolfgang Freude sei besonderer Dank für zahlreiche fruchtbare fachliche Diskussionen und die kritische Durchsicht meiner Manuskripte, einschließlich dieser Dissertation. Sein breites fachliches und linguistisches Verständnis, seine Ruhe und Geduld sowie seine Liebe zum Detail haben die Qualität der Arbeit wesentlich geprägt.

Prof. Christian Koos, dem derzeitigen Institutsleiter, danke ich für die Möglichkeit, meine Forschung auf dem Gebiet der Siliziummodulatoren auch nach Abschluss der Förderprojekte MISTRAL und SOFI weiterzuführen. Danke für die Möglichkeit, neue Kooperationen zu starten, wissenschaftliche Vorträge rund um den Globus zu halten und für die offenen Gespräche.

Viele Ergebnisse dieser Arbeit wurden erst durch die erfolgreiche Zusammenarbeit mit externen Partnern im Rahmen von Förderprojekten ermöglicht. Prof. Wim Bogaerts und Dr. Hui Yu von der Photonics Research Group der Universität Ghent - IMEC danke ich für die Herstellung der Siliziummodulatoren und für ihr Engagement für den Erfolg des Projektes SOFI. Ich danke Dr. Raluca Dinu, GigOptix Bothell, Washington, für die Bereitstellung von elektro-optischem Material. Ich danke Prof. Heinrich Kurz, Dr. Thorsten

Wahlbrink, Dr. Jens Bolten, Dr. Michael Waldow sowie Dr. Matthias Karl von der AMO GmbH, Aachen, für die Herstellung von Teststrukturen und Siliziummodulatoren im Verbundprojekt MISTRAL, sowie für die Bereitstellung der Elektronenmikroskopaufnahme des Schlitzwellenleiters auf dem Cover dieser Dissertation. Spezieller Dank gilt Thorsten Wahlbrink für die intensive Zusammenarbeit und seine stets freundliche und positive Art.

Prof. Larry Dalton und Delwin Elder von der Dalton Research Group der University of Washington, Seattle, danke ich für die Bereitstellung der neuesten elektro-optischen Materialen, die wesentliche zum Erfolg der Arbeit beigetragen haben.

Dem Team des Centrums für funktionelle Nanostrukturen (CFN), darunter Prof. Dr. Gernot Goll, Dr. Silvia Diewald, Stefan Kühn, Patrice Brenner und Heinrich Reimer danke ich für die Bereitstellung der notwendigen Reinraumtechnologie und für technologische Unterstützung. Insbesondere danke ich Dr. Silvia Diewald und Stefan Kühn für das Schreiben zahlreicher Maskenlayouts.

Für die großartige Zusammenarbeit danke ich meinen Kollegen am IPQ. Es war ein Vergnügen, mit euch zu arbeiten! Danke auch für eure Bereitschaft, zusammen die Kunst des Kickerspielens zu erlernen. Die gemeinsam entwickelte Leidenschaft für dieses Spielgerät hat so manche Kaffeepause versüßt, den Alltag „entfrustet“ und zu vielen kreativen Diskussionen geführt.

Ich danke „Chief-Design Engineer“ Dr. Dietmar Korn für die gelungene Zusammenarbeit auf dem Gebiet der SOH Modulatoren. Unsere Arbeitsteilung hat wesentlich dazu beigetragen, dass SOH Modulatoren heute nicht mehr als Forschungskuriosität gelten, sondern als ernstzunehmende Alternative zu kommerziellen Bauteilen. Dietmar, ich danke dir insbesondere für das projektübergreifende, mühevolle Erstellen von PIC-Layouts, welche du trotz zahlreicher „Software-bugs“ hervorragend gemeistert hat.

Ein echter Glücksfall war die Zusammenarbeit mit Dr. Sebastian Köber. Seine Erfahrung auf dem Gebiet elektro-optisch aktiver organischer Materialien sowie der neue Blickwinkel aus Sicht eines Chemikers haben mein Verständnis für nichtlineare Optik auf molekularer Ebene deutlich erweitert. Vielen Dank Sebastian, du hast sehr dazu beigetragen, dass mein letztes Promotionsjahr auch zu meinem erfolgreichsten wurde. Danke auch für viele kurzweilige Unterhaltungen und für die Einführung des „Kannentags“.

Bei Dr. Sven Schüle, meinem ehemaligen Kollegen vom Institut für Mikrostrukturtechnik (IMT, KIT), bedanke ich mich herzlich für die gute institutsübergreifende Zusammenarbeit im Projekt MISTRAL. Die gemeinsame Entwicklung und Demonstration unterschiedlicher „Pigtailing"-Konzepte, größtenteils auf Basis von LIGA-Technologie, hat mir die Röntgenlithographie verständlich gemacht. Dr. Martin Börner vom IMT danke ich für die Organisation und Planung der Herstellung der direktlithographischen Strukturen in diesem Projekt.

Ich danke meinen Bürokollegen Nicole Lindenmann und Tobias Hoose für die angenehme Büroatmosphäre und für die Einbindung in zahlreiche interessante fachliche Diskussionen über „photonic wire bonds".

Meinen ehemaligen Kollegen Dr. Luca Alloatti und Martin Moch danke ich für viele fruchtbare Diskussionen auf dem Gebiet der Silizium-Photonik.

Meinem Kollegen Claudius Weimann danke ich für die Freundschaft seit unserem gemeinsamen Physikstudium. Mir bleiben viele schöne Erinnerungen an gemeinsam Erlebtes, darunter gewisse Junggesellenabschiede und ein St. Patrick's Day in New York.

Für das Engagement im Lehrbetrieb der Halbleiterbauelemente sowie bei der Erarbeitung von Prüfungsaufgaben danke ich meinen Kollegen Sascha Mühlbrandt, Dr. Moritz Röger, Simon Schneider, Matthias Lauermann und Jörg Pfeifle.

Meinen derzeitigen und ehemaligen Kollegen Philipp Schindler, Stefan Wolf, Dr. René Schmogrow, Dr. Swen König, Djorn Karnick, Alexandra Ludwig, Alexander Nesic, Juned Nassir Kemal, Kira Köhnle, Sandeep Ummethala, Muhammad Rodlin Billah, Dr. Sean O'Duill, Dr. Jingshi Li, Argishti Melikyan, Dr. René Bonk, Dr. Thomas Vallaitis, Dr. David Hillerkuß und Dr. Marcus Winter danke ich für die angenehme Gruppenatmosphäre, gemeinsame Kaffee- und Kickerpausen, sowie für gesellige Grillabende auf dem Institutsdach.

Ich danke meinen „Elite-Studenten", darunter Wolfgang Heni, Markus Wößner, Moritz Baier, Ivan Lazar Bundalo und Christoph Reustle für die erfolgreichen Arbeiten am SOH Modulator, welche zu zahlreichen Veröffentlichungen führten, Claudia Hößbacher und Gabriele Navickaite für die fleißige Reinraumarbeit und nass-chemische Herstellung von Silizium Wellenleitern, und ich danke Biying Fu und Padmini Panguluri für die Simulation von Strukturen zur Faser-Chip Kopplung.

Mein herzlicher Dank gilt unserem Sekretariats-Team Bernadette Lehmann, Andrea Riemensperger und Angelika Olbrich, die beispielhaftes „Multi-tasking" praktizierten.

Ein besonderer Dank geht an das Techniker-Team des IPQ, welches sich durch Flexibilität, Kreativität und sehr gutes fachliches Verständnis auszeichnet. Oft aktiv in den Gegenstand der Forschung eingebunden, habt ihr durch eure Erfahrung die Lösung vieler Probleme erst möglich gemacht. Ich danke unserem Feinmechaniker Oswald Speck, dem „Mann mit den ruhigen Händen", für die gute Zusammenarbeit, das gekonnte Schleifen von gewinkelten Glasfasern sowie für die Versorgung mit Frühstücksbrezeln. Ich bedanke mich bei Martin Winkler und David Guder für die Unterstützung und schnelle Problemlösung bei allen Elektronik- und EDV-Belangen. Hans Bürger, Marco Hummel, Werner Höhne und Manfred Hirsch von unserer Mechanikwerkstatt danke ich für die Herstellung zahlreicher Mechanikteile für meine Charakterisierungsaufbauten sowie für die Herstellung der Polungskammer. Ich weiß die Qualität der Arbeit sowie das Einbringen eigener Ideen und Verbesserungsvorschläge sehr zu schätzen! Ich danke Florian Rupp für die Ausübung vielfältiger Aufgaben, wie der Bedienung der Wafersäge, des Elektronenmikroskops und der Arbeit im Lackraum.

Zuletzt gilt mein herzlicher Dank meinen Eltern Elly und Ewald sowie meiner Schwester Nelly. Ganz gleich, was die vergangenen vier Jahre mit sich gebracht haben, ihr Rückhalt und ihre Unterstützung waren mir sicher.

List of Publications

Journal Publications

[J1] T. Alasaarela, D. Korn, L. Alloatti, A. Säynätjoki, A. Tervonen, **R. Palmer**, J. Leuthold, W. Freude, and S. Honkanen, "Reduced propagation loss in silicon strip and slot waveguides coated by atomic layer deposition," *Opt. Expres*, vol. 19, no. 12, pp. 11529–11538, Jun. 2011.

[J2] L. Alloatti, D. Korn, **R. Palmer**, D. Hillerkuss, J. Li, A. Barklund, R. Dinu, J. Wieland, M. Fournier, J. Fedeli, H. Yu, W. Bogaerts, P. Dumon, R. Baets, C. Koos, W. Freude, and J. Leuthold, "42.7 Gbit/s electro-optic modulator in silicon technology," *Opt. Express*, vol. 19, no. 12, pp. 11841–11851, Jun. 2011.

[J3] A. Säynätjoki, L. Karvonen, T. Alasaarela, D. Korn, L. Alloatti, A. Tervonen, **R. Palmer**, C. Koos, J. Leuthold, W. Freude, and S. Honkanen, "Loss reduction of silicon slot waveguides with ALD grown thin films," *Proc. SPIE*, vol. 8266. p. 82660D–82660D–7, 2012.

[J4] **R. Palmer**, L. Alloatti, D. Korn, W. Heni, P. C. Schindler, J. Bolten, M. Karl, M. Waldow, T. Wahlbrink, W. Freude, C. Koos, and J. Leuthold, "Low-Loss Silicon Strip-to-Slot Mode Converters," *IEEE Photonics J.*, vol. 5(1), 2013.

[J5] **R. Palmer**, L. Alloatti, D. Korn, P. C. Schindler, R. Schmogrow, W. Heni, S. Koenig, J. Bolten, T. Wahlbrink, M. Waldow, H. Yu, W. Bogaerts, P. Verheyen, G. Lepage, M. Pantouvaki, J. Van Campenhout, P. Absil, R. Dinu, W. Freude, C. Koos, and J. Leuthold, "Silicon-Organic Hybrid MZI Modulator Generating OOK, BPSK and 8-ASK Signals for Up to 84 Gbit/s," *IEEE Photonics J.*, vol. 5, no. 2, pp. 6600907–6600907, Apr. 2013.

[J6] D. Korn, **R. Palmer**, H. Yu, P. C. Schindler, L. Alloatti, M. Baier, R. Schmogrow, W. Bogaerts, S. K. Selvaraja, G. Lepage, M. Pantouvaki, J. M. D. Wouters, P. Verheyen, J. Van Campenhout, B. Chen, R. Baets, P. Absil, R. Dinu, C. Koos, W. Freude, and J. Leuthold, "Silicon-organic hybrid (SOH) IQ modulator using the linear electro-optic effect for transmitting 16QAM at 112 Gbit/s," *Opt. Express*, vol. 21, no. 11, pp. 13219–13227, Jun. 2013.

[J7] **R. Palmer**, L. Alloatti, D. Korn, P. C. Schindler, M. Baier, J. Bolten, T. Wahlbrink, M. Waldow, R. Dinu, W. Freude, C. Koos, and J. Leuthold, "Low Power Mach–Zehnder Modulator in Silicon-Organic Hybrid Technology," *IEEE Phot. Technol. Lett.*, vol. 25, no. 13, pp. 1226–1229, Jul. 2013.

[J8] J. Leuthold, C. Koos, W. Freude, L. Alloatti, **R. Palmer**, D. Korn, J. Pfeifle, M. Lauermann, R. Dinu, S. Wehrli, M. Jazbinsek, P. Gunter, M. Waldow, T. Wahlbrink, J. Bolten, H. Kurz, M. Fournier, J.-M. Fedeli, H. Yu, and W. Bogaerts, "Silicon-Organic Hybrid Electro-Optical Devices," *IEEE J. Sel. Top. Quantum Electron.*, vol. 19, no. 6, pp. 3401413–3401413, Nov. 2013.

[J9] S. Koenig, D. Lopez-Diaz, J. Antes, F. Boes, R. Henneberger, A. Leuther, A. Tessmann, R. Schmogrow, D. Hillerkuss, **R. Palmer**, T. Zwick, C. Koos, W. Freude, O. Ambacher, J. Leuthold, and I. Kallfass, "Wireless sub-THz communication system with high data rate," *Nat Phot.*, vol. 7, no. 12, pp. 977–981, Dec. 2013.

[J10] P. C. Schindler, D. Korn, C. Stamatiadis, M. F. O'Keefe, L. Stampoulidis, R. Schmogrow, P. Zakynthinos, **R. Palmer**, N. Cameron, Y. Zhou, R. G. Walker, E. Kehayas, S. Ben-Ezra, I. Tomkos, L. Zimmermann, K. Petermann, W. Freude, C. Koos, and J. Leuthold, "Monolithic GaAs Electro-Optic IQ Modulator Demonstrated at 150 Gbit/s With 64QAM," *Light. Technol. J.*, vol. 32, no. 4, pp. 760–765, Feb. 2014.

[J11] C. Weimann, P. C. Schindler, **R. Palmer**, S. Wolf, D. Bekele, D. Korn, J. Pfeifle, S. Koeber, R. Schmogrow, L. Alloatti, D. Elder, H. Yu, W. Bogaerts, L. R. Dalton, W. Freude, J. Leuthold, and C. Koos, "Silicon-organic hybrid (SOH) frequency comb sources for terabit/s data transmission," *Opt. Express*, vol. 22, no. 3, pp. 3629–3637, Feb. 2014.

[J12] A. Melikyan, L. Alloatti, A. Muslija, D. Hillerkuss, P. C. Schindler, J. Li, **R. Palmer**, D. Korn, S. Muehlbrandt, D. Van Thourhout, B. Chen, R. Dinu, M. Sommer, C. Koos, M. Kohl, W. Freude, and J. Leuthold, "High-speed plasmonic phase modulators," *Nat. Photonics*, vol. 8, no. 3, pp. 229–233, Feb. 2014.

[J13] **R. Palmer**, S. Koeber, D. Elder, M. Woessner, W. Heni, D. Korn, M. Lauermann, W. Bogaerts, L. Dalton, W. Freude, J. Leuthold, and

C. Koos, "High-Speed, Low Drive-Voltage Silicon-Organic Hybrid Modulator based on a Binary-Chromophore Electro-Optic Material," *J. Light. Technol.*, May 2014.

[J14] M. Lauermann, **R. Palmer**, S. Koeber, P. C. Schindler, D. Korn, T. Wahlbrink, J. Bolten, M. Waldow, D. L. Elder, L. R. Dalton, J. Leuthold, W. Freude, and C. Koos, "Low-power silicon-organic hybrid (SOH) modulators for advanced modulation formats.," *Opt. Express*, vol. 22, no. 24, pp. 29927–36, Dec. 2014.

[J15] S. Koeber, **R. Palmer**, M. Lauermann, W. Heni, D. L. Elder, D. Korn, M. Woessner, L. Alloatti, S. Koenig, P. C. Schindler, H. Yu, W. Bogaerts, L. R. Dalton, W. Freude, J. Leuthold, and C. Koos, "Femtojoule electro-optic modulation using a silicon–organic hybrid device," *Nat. Light Sci. Appl.*, vol. 4, no. 2, p. e255, Feb. 2015.

[J16] M. Lauermann, S. Wolf, P. Schindler, **R. Palmer**, S. Koeber, D. Korn, L. Alloatt, T. Wahlbrink, J. Bolten, M. Waldow, M. Koenigsmann, M. Kohler, D. Malsam, D. Elder, P. Johnston, N. Phillips-Sylvain, P. Sullivan, L. Dalton, J. Leuthold, W. Freude, and C. Koos, "40 GBd 16QAM Signaling at 160 Gbit/s in a Silicon-Organic Hybrid (SOH) Modulator," *J. Light. Technol.*, 2015.

[J17] S. Wolf, M. Lauermann, P. Schindler, G. Ronniger, K. Geistert, **R. Palmer**, S. Koeber, W. Bogaerts, J. Leuthold, W. Freude, and C. Koos, "DAC-less Amplifier-less Generation and Transmission of QAM Signals Using Sub-Volt Silicon-organic Hybrid Modulators," *J. Light. Technol.*, 2015.

Conference Proceedings

[C1] W. Freude, J. Leuthold, L. Alloatti, T. Vallaitis, D. Korn, **R. Palmer**, C. Koos, J.-M. Brosi, P. Dumon, R. Baets, M. L. Scimeca, I. Biaggio, B. Breiten, F. Diederich, A. Barklund, R. Dinu, and J. . Wieland, "100 Gbit/s electro-optic modulator and 56 Gbit/s wavelength converter for DQPSK data in silicon-organic hybrid (SOH) technology," in *Proc. IEEE Photonics Society* (formerly: Lasers and Electro-Optics Society) Summer Topicals 2010, Paper WB2.1, 2010.

[C2] L. Alloatti, D. Korn, D. Hillerkuss, T. Vallaitis, J. Li, R. Bonk, **R. Palmer**, T. Schellinger, A. Barklund, R. Dinu, J. Wieland, M. Fournier, J. Fedeli, P. Dumon, R. Baets, C. Koos, W. Freude, and J. Leuthold, "40 Gbit/s silicon-organic hybrid (SOH) phase modulator," in *36th European Conference and Exhibition on Optical Communication (ECOC 2010)*, 2010.

[C3] L. Alloatti, D. Korn, D. Hillerkuss, T. Vallaitis, J. Li, R. Bonk, **R. Palmer**, T. Schellinger, C. Koos, W. Freude, J. Leuthold, A. Barklund, R. Dinu, J. Wieland, M. Fournier, J. Fedeli, W. Bogaerts, D. P., and R. Baets, "Silicon high-speed electro-optic modulator," in *Proc. Group IV Photonics 2010*, 2010.

[C4] W. Freude, L. Alloatti, T. Vallaitis, D. Korn, D. Hillerkuss, R. Bonk, **R. Palmer**, J. Li, T. Schellinger, M. Fournier, J. Fedeli, W. Bogaerts, P. Dumon, R. Baets, A. Barklund, R. Dinu, J. Wieland, M. L. Scimeca, I. Biaggio, B. Breiten, F. Diederich, C. Koss, and J. Leuthold, "High-speed signal processing with silicon-organic hybrid devices," in *Journal of the european optical society*, vol. 5, 2010

[C5] **R. Palmer**, L. Alloatti, D. Korn, M. Moosmann, K. Huska, U. Lemmer, D. Gerthsen, T. Schimmel, W. Freude, C. Koos, and J. Leuthold, "Smooth and ultra-precise silicon nanowires fabricated by conventional optical lithography," in *CLEO 2011 - Laser Applications to Photonic Applications*, paper CThZ1, 2011.

[C6] C. Koos, L. Alloatti, D. Korn, **R. Palmer**, D. Hillerkuss, J. Li, A. Barklund, R. Dinu, J. Wieland, M. Fournier, J.-M. Fedeli, H. Yu, W. Bogaerts, P. Dumon, R. Baets, W. Freude, and J. Leuthhold, "Silicon-Organic Hybrid (SOH) Electro-Optical Devices," in *Advanced Photonics*, paper. IWF1, 2011.

[C7] J. Leuthold, W. Freude, C. Koos, L. Alloatti, D. Korn, **R. Palmer**, and J. M. Brosi, "Slotted Photonic Crystal Slow Light Modulators," in *Advanced Photonics*, paper. JTuA3; 2011.

[C8] C. Koos, L. Alloatti, D. Korn, **R. Palmer**, T. Vallaitis, R. Bonk, D. Hillerkuss, J. Li, W. Bogaerts, P. Dumon, R. Baets, M. L. Scimeca, I. Biaggio, A. Barklund, R. Dinu, J. Wieland, M. Fournier, J. Fedeli, W. Freude, and J. Leuthold, "Silicon nanophotonics and silicon-organic hybrid (SOH) integration," in *General Assembly and Scientific Symposium 2011 XXXth URSI*, 2011.

[C9] N. Lindenmann, G. Balthasar, **R. Palmer**, S. Schuele, J. Leuthold, W. Freude, and C. Koos, "Photonic wire bonding for single-mode chip-to-chip interconnects," in *Group IV Photonics (GFP 2011)*, pp. 380–382, 2011.

[C10] P. C. Schindler, H. Fischer, H. Haisch, N. Lindenmann, L. Alloatti, **R. Palmer**, D. Korn, D. Hillerkuss, R. Schmogrow, A. Gerster, W. Freude, J. Leuthold, and C. Koos, "Optical Technologies for Off-Chip and On-Chip Interconnects," in *DesignCon 2012*, 2012.

[C11] A. Säynätjoki, L. Karvonen, T. Alasaarela, D. Korn, L. Alloatti, A. Tervonen, **R. Palmer**, C. Koos, J. Leuthold, W. Freude, and S. Honkanen, "Loss reduction of silicon slot waveguides with ALD grown thin films," *Proc. SPIE*, vol. 8266. p. 82660D–82660D–7, 2012.

[C12] W. Freude, L. Alloatti, A. Melikyan, **R. Palmer**, D. Korn, N. Lindenmann, T. Vallaitis, D. Hillerkuss, J. Li, A. Barklund, R. Dinu, J. Wieland, M. Fournier, J.-M. Fedeli, S. Walheim, P. Leufke, S. Ulrich, J. Ye, P. Vincze, H. Hahn, H. Yu, W. Bogaerts, P. Dumon, R. Baets, B. Breiten, F. Diederich, M. T. Beels, I. Biaggio, T. Schimmel, C. Koos, and J. Leuthold, "Nonlinear Optics on the Silicon Platform," in *Optical Fiber Communication Conference (OFC 2012)*, paper OTh3H.6, 2012.

[C13] C. Koos, J. Leuthold, W. Freude, L. Alloatti, D. Korn, N. Lindenmann, **R. Palmer**, J. Pfeifle, D. Hillerkuss, and R. Schmogrow, "Silicon Nanophotonics and Photonic Wire Bonding: Technologies for Terabit/s Interconnects," in *6th ITG Workshop on Photonic Integration, Laser Optics Berlin*, 2012.

[C14] D. Korn, M. Lauermann, P. Appel, L. Alloatti, **R. Palmer**, W. Freude, J. Leuthold, and C. Koos, "First Silicon-Organic Hybrid Laser at

Telecommunication Wavelength," in *CLEO: Science and Innovations (CLEO 2012)*, paper CTu2I.1, 2012.

[C15] **R. Palmer**, L. Alloatti, D. Korn, W. Heni, P. Schindler, J. Bolten, M. Karl, M. Waldow, T. Wahlbrink, W. Freude, C. Koos, and J. Leuthold, "Highly Efficient Strip-to-Slot Mode Converters," *CLEO: Science and Innovations (CLEO 2012)*, San Jose, 2012.

[C16] D. Korn, L. Alloatti, M. Lauermann, J. Pfeifle, **R. Palmer**, P. C. Schindler, W. Freude, C. Koos, J. Leuthold, H. Yu, W. Bogaerts, K. Komorowska, R. Baets, J. Van Campenhout, P. Verheyen, J. Wouters, M. Moelants, P. Absil, A. Secchi, M. Dispenza, M. Jazbinsek, P. Gunter, S. Wehrli, M. Bossard, P. Zakynthinos, and I. Tomkos, "Silicon-organic hybrid fabrication platform for integrated circuits," in *14th International Conference on Transparent Optical Networks (ICTON 2012)*, 2012.

[C17] C. Koos., J. Leuthold, W. Freude, L. Alloatti, D. Korn, **R. Palmer**, M. Lauermann, N. Lindenmann, S. Koeber, J. Pfeifle, P. C. Schindler, D. Hillerkuss, and R. Schmogrow, "Silicon-Organic Hybrid Integration and Photonic Wire Bonding: Technologies for Terabit/s Interconnects," in *Joint Symposium on Opto- and Microelectronic Devices and Circuits (SODC 2012)*, Hangzhou, China, 2012.

[C18] J. Leuthold, C. Koos, W. Freude, L. Alloatti, **R. Palmer**, D. Korn, L. Pfeifle, and L. Lauermann, "Silicon-Organic Hybrid - a path towards active silicon photonic devices," in *Proc. EOS Annual Meeting (EOSAM 2012)*, 2012.

[C19] J. Leuthold, et al., "Ultracompact CMOS-compatible Modulators," in *Frontiers in Optics 2012/Laser Science XXVIII*, paper FTu4A.1, 2012.

[C20] L. Alloatti, D. Korn, J. Pfeifle, **R. Palmer**, S. Koeber, M. Baier, R. Schmogrow, S. Diebold, P. Pahl, T. Zwick, H. Yu, W. Bogaerts, R. Baets, M. Fournier, J. Fedeli, R. Dinu, C. Koos, W. Freude, and J. Leuthold, "Silicon-organic hybrid devices," *Proc. SPIE*, vol. 8629, p. 86290P–86290P–12, 2013.

[C21] **R. Palmer**, L. Alloatti, D. Korn, P. C. Schindler, R. M. Schmogrow, M. Baier, S. Koenig, D. Hillerkuss, J. Bolten, T. Wahlbrink, M. Waldow, R. Dinu, W. Freude, C. Koos, and J. Leuthold, "Silicon-Organic Hybrid (SOH) Modulator Generating up to 84 Gbit/s

BPSK and M-ASK Signals," in *Optical Fiber Communication Conference (OFC 2013)*, paper OW4J.6, 2013, (Corning Outstanding Student Paper Award).

[C22] S. Koenig, F. Boes, D. Lopez-Diaz, J. Antes, R. Henneberger, R. M. Schmogrow, D. Hillerkuss, **R. Palmer**, T. Zwick, C. Koos, W. Freude, O. Ambacher, I. Kallfass, and J. Leuthold, "100 Gbit/s Wireless Link with mm-Wave Photonics," in *Optical Fiber Communication Conference (OFC 2013)*, paper PDP5B.4, 2013.

[C23] D. Korn, P. C. Schindler, C. Stamatiadis, M. F. O'Keefe, L. Stampoulidis, R. M. Schmogrow, P. Zakynthinos, **R. Palmer**, N. Cameron, Y. Zhou, R. G. Walker, E. Kehayas, I. Tomkos, L. Zimmermann, K. Petermann, W. Freude, C. Koos, and J. Leuthold, "First Monolithic GaAs IQ Electro-optic Modulator, Demonstrated at 150 Gbit/s with 64-QAM," in *Optical Fiber Communication Conference (OFC 2013)*, paper PDP5C.4, 2013.

[C24] D. Korn, **R. Palmer**, H. Yu, P. C. Schindler, L. Alloatti, M. Baier, R. Schmogrow, W. Bogaerts, S. Selvaraja, G. Lepage, M. Pantouvaki, J. Wouters, P. Verheyen, J. Van Campenhout, P. Absil, R. Baets, R. Dinu, C. Koos, W. Freude, and J. Leuthold, "Silicon-organic hybrid (SOH) IQ modulator for 16QAM at 112 Gbit/s," in *Conference on Lasers and Electro-Optics Europe (CLEO-Europe/IQEC 2013)*, Munich, Germany, 2013.

[C25] A. Melikyan, L. Alloatti, A. Muslija, D. Hillerkuss, P. Schindler, J. Li, **R. Palmer**, D. Korn, S. Muehlbrandt, D. Van Thourhout, B. Chen, R. Dinu, M. Sommer, C. Koos, M. Kohl, W. Freude, and J. Leuthold, "Surface Plasmon Polariton High-Speed Modulator," in *Conference on Lasers and Electro-Optics (CLEO 2013),* PDP paper CTh5D.2, 2013.

[C26] C. Koos, J. Leuthold, W. Freude, L. Alloatti, **R. Palmer**, D. Korn, J. Pfeifle, P. C. Schindler, and M. Lauermann, "Silicon-organic hybrid (SOH) technology: A platform for efficient electro-optical devices," in *International Conference on Optical MEMS and Nanophotonics (OMN 2013)*, pp. 85–86, 2013.

[C27] C. Weimann, S. Wolf, D. Korn, **R. Palmer**, S. Koeber, R. Schmogrow, P. C. Schindler, L. Alloatti, A. Ludwig, W. Heni, D. Bekele, D. L. Elder, H. Yu, W. Bogaerts, L. R. Dalton, W. Freude, J. Leuthold, and C. Koos,

"Silicon-organic hybrid (SOH) frequency comb source for data transmission at 784 Gbit/s," in *39th European Conference and Exhibition on Optical Communication (ECOC 2013)*, 2013.

[C28] **R. Palmer**, S. Koeber, W. Heni, D. L. Elder, D. Korn, H. Yu, L. Alloatti, S. Koenig, P. C. Schindler, W. Bogaerts, M. Pantouvaki, G. Lepage, P. Verheyen, J. Van Campenhout, P. Absil, R. Baets, L. R. Dalton, W. Freude, J. Leuthold, and C. Koos, "High-Speed Silicon-Organic Hybrid (SOH) Modulator with 1.6 fJ/bit and 180 pm/V In-Device Nonlinearity," in *39th European Conference and Exhibition on Optical Communication (ECOC 2013)*, pp. 510–512, 2013 [invited] (ADVA Best Student Paper Award).

[C29] W. Freude, L. Alloatti, D. Korn, M. Lauermann, A. Melikyan, **R. Palmer**, J. Pfeifle, P. C. Schindler, C. Weimann, R. Dinu, J. Bolten, T. Wahlbrink, M. Waldow, S. Walheim, P. Leufke, S. Ulrich, J. Ye, P. Vincze, H. Hahn, H. Yu, W. Bogaerts, V. Brasch, T. Herr, R. Holzwarth, K. Hartinger, C. Stamatiadis, M. O'Keefe, L. Stampoulidis, L. Zimmermann, R. Baets, T. Schimmel, I. Tomkos, K. Petermann, T. J. Kippenberg, C. G. Koos, and J. Leuthold, "Nonlinear Nano-Photonics," in *Frontiers in Optics 2013*, paper FTu4C.6, 2013.

[C30] C. Weimann, P. Schindler, D. Bekele, **R. Palmer**, D. Korn, J. Pfeifle, S. Koeber, R. Schmogrow, L. Alloatti, D. L. Elder, H. Yu, W. Bogaerts, L. R. Dalton, W. Freude, J. Leuthold, and C. Koos, "Data Transmission at Terabit/s Data Rates Using Silicon-Organic Hybrid (SOH) Frequency Combs," in *Optical Fiber Communication Conference (OFC 2014)*, paper Th4I.2, 2014.

[C31] M. Lauermann, C. Weimann, A. Knopf, D. L. Elder, W. Heni, **R. Palmer**, D. Korn, P. Schindler, S. Koeber, L. Alloatti, H. Yu, W. Bogaerts, L. R. Dalton, C. Rembe, J. Leuthold, W. Freude, and C. Koos, "Integrated Silicon-Organic Hybrid (SOH) Frequency Shifter," in *Optical Fiber Communication Conference (OFC 2014)*, paper Tu2A.1, 2014.

[C32] **R. Palmer**, S. Koeber, M. Woessner, D. L. Elder, W. Heni, D. Korn, H. Yu, M. Lauermann, W. Bogaerts, L. R. Dalton, W. Freude, J. Leuthold, and C. Koos, "High-Speed Silicon-Organic Hybrid (SOH) Modulators with 230 pm/V Electro-Optic Coefficient Using Advanced Materials," in

Optical Fiber Communication Conference (OFC 2014), paper M3G.4, 2014.

[C33] K. Kohnle, M. Lauermann, A. Bacher, M.-K. Nees, S. Muehlbrandt, **R. Palmer**, A. Muslija, P.-J. Jakobs, W. Freude, and C. Koos, "Silicon-Organic Hybrid Phase Modulator Based on Sub-Wavelength Grating Waveguides," in *Micro and Nano Engineering (MNE2014)*, Lausanne, Switzerland, paper 8127, Sept. 2014.

[C34] A. Melikyan, L. Alloatti, A. Muslija, D. Hillerkuss, P. C. Schindler, J. Li, **R. Palmer**, D. Korn, N. Lindenmann, S. Muehlbrandt, S. Walheim, P. Vincze, P. M. Leufke, S. Ulrich, J. Ye, D. Van Thourhout, B. Chen, R. Dinu, M. Sommer, H. Hahn, T. Schimmel, C. Koos, M. Kohl, W. Freude, and J. Leuthold, "High-speed Plasmonic Modulators," in *Advanced Photonics for Communications*, paper IT2A.6, 2014.

[C35] J. Leuthold, A. Melikyan, L. Alloatti, D. Korn, **R. Palmer**, D. Hillerkuss, M. Lauermann, P. C. Schindler, B. Chen, R. Dinu, D. L. Elder, L. R. Dalton, C. Koos, M. Kohl, W. Freude, and C. Hafner, "From silicon-organic hybrid to plasmonic modulation," in *European Conference on Optical Communication (ECOC 2014)*, 2014.

[C36] **R. Palmer**, S. Koeber, M. Woessner, W. Heni, D. L. Elder, D. Korn, C. Weimann, P. C. Schindler, D. Bekele, L. Alloatti, W. Bogaerts, L. R. Dalton, J. Leuthold, W. Freude, and C. Koos, "Progress in silicon-organic hybrid (SOH) integration," in *11th International Conference on Group IV Photonics (GFP 2014)*, 2014, [invited].

[C37] C. Koos, W. Freude, T. J. Kippenberg, J. Leuthold, L. R. Dalton, J. Pfeifle, C. Weimann, M. Lauermann, **R. Palmer**, S. Koeber, S. Wolf, P. Schindler, V. Brasch, and D. Elder, "Terabit/s optical transmission using chip-scale frequency comb sources," in *European Conference on Optical Communication (ECOC 2014)*, 2014.

[C38] P. C. Schindler, M. Lauermann, S. Wolf, D. Korn, **R. Palmer**, S. Koeber, W. Heni, A. Ludwig, R. Schmogrow, D. L. Elder, L. R. Dalton, W. Bogaerts, H. Yu, W. Freude, J. Leuthold, and C. Koos, "Ultra-short silicon-organic hybrid (SOH) modulator for bidirectional polarization-independent operation," in *European Conference on Optical Communication (ECOC 2014)*, 2014.

[C39] S. Wolf, P. C. Schindler, G. Ronniger, M. Lauermann, **R. Palmer**, S. Koeber, D. Korn, W. Bogaerts, J. Leuthold, W. Freude, and C. Koos, "10 GBd SOH modulator directly driven by an FPGA without electrical amplification," in *European Conference on Optical Communication (ECOC 2014)*, 2014.

[C40] S. Wolf, M. Lauermann, G. Ronniger, P. C. Schindler, **R. Palmer**, S. Koeber, W. Bogaerts, J. Leuthold, W. Freude, and C. Koos, "DAC-less and amplifier-less generation and transmission of 16QAM signals using a sub-volt silicon photonic modulator," in *European Conference on Optical Communication (ECOC 2014)*, 2014.

[C41] M. Lauermann, P. C. Schindler, S. Wolf, **R. Palmer**, S. Koeber, D. Korn, L. Alloatti, T. Wahlbrink, J. Bolten, M. Waldow, M. Koenigsmann, M. Kohler, D. Malsam, D. L. Elder, P. V. Johnston, N. Phillips-Sylvain, P. A. Sullivan, L. R. Dalton, J. Leuthold, W. Freude, and C. Koos, "40 GBd 16QAM modulation at 160 Gbit/s in a silicon-organic hybrid (SOH) modulator," in *European Conference on Optical Communication (ECOC 2014)*, 2014.

[C42] S. Koenig, D. Lopez-Diaz, J. Antes, F. Boes, R. Henneberger, A. Leuther, A. Tessmann, R. Schmogrow, D. Hillerkuss, **R. Palmer**, T. Zwick, C. Koos, W. Freude, O. Ambacher, J. Leuthold, and I. Kallfass, "Wireless sub-THz communication system with high data rate enabled by RF photonics and active MMIC technology," in *IEEE Photonics Conference*, pp. 414–415, 2014.

[C43] S. Koenig, D. Lopez-Diaz, J. Antes, F. Boes, R. Henneberger, A. Leuther, A. Tessmann, R. Schmogrow, D. Hillerkuss, **R. Palmer**, T. Zwick, C. Koos, W. Freude, O. Ambacher, J. Leuthold, and I. Kallfass, "Wireless sub-THz communication system with high data rate enabled by RF photonics and active MMIC technology," in *IEEE Photonics Conference*, pp. 414–415, 2014.

[C44] **R. Palmer**, M. Lauermann, and C. Koos, "Silicon-Organic Hybrid (SOH) Devices for Applications in Optical Communications, Metrology, and Sensing," in *115th Annual DGaO Conference (DGaO'14)*, June 10 - 14, Karlsruhe, Germany , talk S8, Deutsche Gesellschaft für angewandte Optik e.V. (DGaO), 2014.

[C45] C. Koos, W. Freude, **R. Palmer**, M. Lauermann, S. Kober, J. Pfeifle, C. Weimann, P. C. Schindler, N. Lindenmann, T. Hoose, and M. R. Billah, "Silicon Photonics, Hybrid Integration, and Frequency Combs: Technologies for Terabit/s Communications, Teratronics, and Optical Metrology," in *115th Annual DGaO Conference (DGaO'14)*, June 10 - 14, Karlsruhe, Germany , talk H1 Deutsche Gesellschaft für angewandte Optik e.V. (DGaO), 2014.

[C46] M. Lauermann, S. Wolf, **R. Palmer**, A. Bielik, L. Altenhain, J. Lutz, R. Schmid, T. Wahlbrink, J. Bolten, A. L. Giesecke, W. Freude, and C. Koos, "64 GBd operation of a silicon-organic hybrid modulator at elevated temperature," in *Optical Fiber Communication Conference (OFC 2015)*, Los Angeles (CA), paper Tu2A.5, March 2015.

[C47] C. Koos, J. Leuthold, W. Freude, M. Kohl, L. R. Dalton, W. Bogaerts, A. L. Giesecke, M. Lauermann, A. Melikyan, S. Kober, S. Wolf, S. Weimann, C. Muehlbrandt, K. Kohnle, J. Pfeifle, **R. Palmer**, L. Alloatti, D. L. Elder, T. Wahlbrink, and J. Bolten, "Silicon-Organic Hybrid (SOH) and Plasmonic-Organic Hybrid (POH) Integration," in *Optical Fiber Communication Conference (OFC 2015)*, Los Angeles (CA), paper Tu2A, March 2015, [invited].

Workshop

[W1] **R. Palmer**, C. Koos, "Hybrid Photonic Integration: Enabling Technology for Terabit/s Communications," *ADOPT Winter School 2014*, Romme Alpin, Borlänge, Sweden, 2014.

Curriculum Vitae

Robert Palmer

born on August 31th, 1985
in Pawlodar, Kazakhstan

Nationality: German

Education

06/2010–06/2014	Ph.D. candidate and research associate at the Institute of Photonics and Quantum Electronics at the Karlsruhe Institute of Technology. Field of study: Silicon Photonics
04/2005–05/2010	Studies at the Karlsruhe Institute of Technology, Germany (former University of Karlsruhe) Major: Physics Specialization: Semiconductor-physics, nonlinear optics Thesis: "A New Method to Fabricate Optical Nano-Wire Waveguides" Degree: Dipl.-Phys. (hons.)

School Education

01/2003–03/2005	"Abitur" (grade 1.3, "very good"), Max-Slevogt-Gymnasium, Landau, Germany (general university entrance qualification)
08/2002–01/2003	John-Abbott College/CEGEP, Montreal, Canada
08/1996–06/2002	Max-Slevogt-Gymnasium, Landau, Germany

Honors and Awards

09/2013	ADVA Best Student Paper Award – First prize at the European Conference on Optical Communication (ECOC'13)
03/2013	Corning Outstanding Student Paper Award – First prize at the Optical Fiber Communication Conference (OFC'13)
03/2010	Diploma in physics with honors